Cerebral Blood Flow in Acute Head Injury

The Regulation of Cerebral Blood Flow and Metabolism
During the Acute Phase of Head Injury,
and Its Significance for Therapy

Georg Emil Cold

Acta Neurochirurgica
Supplementum 49

Springer-Verlag Wien New York

Georg Emil Cold, M.D.
Department of Neuroanesthesiology, Århus Kommunehospital, Århus, Denmark

With 16 Figures

Library of Congress Cataloging-in-Publication Data. Cold, G. E. (Georg Emil), 1938– . Cerebral blood flow in acute head injury: the regulation of cerebral blood flow and metabolism during the acute phase of head injury, and its significance for therapy / Georg Emil Cold. p. cm. – (Acta neurochirurgica. Supplementum, ISSN 0065-1419; 49). Includes bibliographical references. ISBN-13:978-3-7091-9103-3 1. Brain damage. 2. Cerebral circulation. 3. Brain – Metabolism. I. Title. II. Series. [DNLM: 1. Brain – blood supply. 2. Cerebrovascular Circulation. 3. Head Injuries – metabolism. 4. Head Injuries – physiopathology. W1 AC8661 no. 49 / WL 302 C688c]. RC387.5.C64 1990. 617.4'81044 – dc20. 90-10379.

ISSN 0065-1419
ISBN-13:978-3-7091-9103-3   e-ISBN-13:978-3-7091-9101-9
DOI: 10.1007/978-3-7091-9101-9

Preface

The present studies were carried out at the Department of Neurosurgery G, Århus University Hospital 1970–1973; at the Department of Anesthesiology, Hvidovre University Hospital, Copenhagen, 1976–1977; and finally at the Department of Neurosurgery GS, Århus University Hospital, 1986–1987. Accordingly, I want to thank the chiefs of these departments, Professor Richard Malmros, Professor Peter Rasmussen, and Jens Buhl in Århus, and Professor Henning Ruben in Copenhagen.

I am deeply indebted to my collaborators in these clinical studies. For help during the first studies, I owe special thanks to Hans Hvid Hansen M.D., Head of the Department of Nuclear Medicine, Finn Tågehøj Jensen B.Sc., and Erna Enevoldsen M.D. Regarding the studies performed in Copenhagen, I am indebted to Mogens Stig Christensen M.D. and Kåre Schmidt M.D. Dr. E. Spencer provided valuable assistance with regard to the linguistic formulation of the review.

Furthermore, I am indebted to my colleagues anaesthesiologists Erland Hansen M.D. and Jørgen Kaalund Jensen, who encouraged me to continue the clinical studies. Finally, I want to express my gratitude to my wife Ida Cold and my sons Jakob and Christian for their patience and understanding during two decades.

Georg Emil Cold

This review is based on the following previously published papers:

I   Cold GE, Jensen FT, Malmros R (1977) The cerebrovascular CO_2 reactivity during the acute phase of brain injury. Acta Anaesthesiol Scand 21: 222–231.

II   Cold GE, Jensen FT, Malmros R (1977) The effects of $PaCO_2$ reduction on regional cerebral blood flow in the acute phase of brain injury. Acta Anaesthesiol Scand 21: 359–367.

III   Cold GE (1978) Cerebral metabolic rate of oxygen ($CMRO_2$) in the acute phase of brain injury. Acta Anaesthesiol Scand 22: 249–256.

IV   Cold GE, Jensen FT (1978) Cerebral autoregulation in unconscious patients with brain injury. Acta Anaesthesiol Scand 22: 270–280.

V   Cold GE, Jensen FT (1980) Cerebral blood flow in the acute phase after head injury. Part I: Correlation to age of the patients, clinical outcome and localization of the injured region. Acta Anaesthesiol Scand 24: 245–251.

VI   Cold GE, Christensen MS, Schmidt K (1981) Effect of two levels of induced hypocapnia on cerebral autoregulation in the acute phase of head injury coma. Acta Anaesthesiol Scand 25: 397–401.

VII   Cold GE (1986) The relationship between cerebral metabolic rate of oxygen and cerebral blood flow in the acute phase of head injury. Acta Anaesthesiol Scand 30: 453–457.

VIII   Cold GE (1989) Does acute hyperventilation provoke cerebral oligaemia in comatose patients after acute head injury? Acta Neurochir (Wien) 96: 100–106.

IX   Cold GE (1989) Measurements of CO_2 reactivity and barbiturate reactivity in patients with severe head injury. Acta Neurochir (Wien) 98: 153–163.

Contents

Listed in Current Contents

Abbreviations

$AVDO_2$	Arterio-venous difference of oxygen
BBB	Blood-brain barrier
BI	Brain injury
CA	Cerebral autoregulation
CBF	Cerebral blood flow
CBV	Cerebral blood volume
$CMRO_2$	Cerebral metabolic rate of oxygen
CPP	Cerebral perfusion pressure
CVP	Central venous pressure
CVR	Cerebral vascular resistance
CPH	Controlled prolonged hyperventilation
CSF	Cerebrospinal fluid
GCS	Glasgow coma score
HI	Head injury
ICP	Intracranial pressure
IH	Intracranial hypertension
MABP	Mean arterial blood pressure
MCAO	Middle cerebral artery occlusion
MR	Magnetic resonance

Introduction

Understanding of the dynamic changes in cerebral blood flow (CBF), cerebral metabolic rate of oxygen (CMRO$_2$), and intracranial pressure (ICP) is of utmost importance in the clinical care of patients with severe head injury (HI). Thus, four principles of treatment in common use in the management of patients with HI (i.e. prolonged artifical hyperventilation, barbiturate sedation, hypothermia and mannitol treatment) are based on the principles for regulation of CBF, CMRO$_2$, and ICP. During the last decade a multitude of studies concerning the dynamic changes in CBF, CMRO$_2$, and ICP have been published. These studies have been supplemented with studies of cerebral autoregulation (CA) and studies of the chemical regulation of CBF, especially CO$_2$ reactivity. Recently the topic has been reviewed (Enevoldsen 1980, Jennett and Teasdale 1981, Enevoldsen 1986, Sundbärg 1988); however, the therapeutical implications of clinical CBF studies have only rarely been discussed. In this review experimental and clinical studies of cerebral circulation and metabolism in severe head injury have been supplemented with studies of intracranial pressure (ICP), biochemical studies of brain tissue and cerebrospinal fluid, and the theoretical implications for therapy are discussed. Electroencephalographic investigations, studies with transcranial doppler technique, studies of magnetic resonance and CT scanning have only been considered if they elucidated the dynamic changes in CBF, ICP or metabolism.

The review consists of five chapters. Chapter 1 considers experimental and human methodology of CBF studies, metabolism and the regulation of CBF. Chapter 2 reviews present knowledge of ischaemic threshold. On the basis of experimental and clinical studies concerning the effects of barbiturates and mannitol on cerebral circulation and metabolism, studies concerning barbiturate treatment and the use of mannitol in patients with head injury are reviewed in chapter 3. Chapter 4 considers experimental and clinical studies of ICP, CBF and metabolism after acute head injury. In chapter 5 human studies of ICP and, biochemical studies of CSF are summarized. Chapter 6 includes a presentation of the author's studies of CBF and metabolism in the acute phase of head injury and the chapter is concluded with a general discussion. A summary concerning the dynamic changes in cerebral circulation and metabolism is elaborated and followed by theoretical considerations concerning the therapeutic effect of prolonged hyperventilation, barbiturate treatment, and treatment with mannitol.

This review is primary addressed to neurosurgeons and neuroanaesthesiologists concerned with the management of patients with severe head injury. Studies of subarachnoid haemorrhage, cerebral tumours and intracerebral haematoma have not been reviewed, unless these investigations were relevant to other aspects in the text. New principles of treatment including Ca^{++} blocking agents and indomethacin have not been reviewed. Nevertheless, the aim is, that the text presented might offer some help in the understanding of the dynamic changes in cerebral circulation and metabolism in neurosurgical and neuroanaesthesiological practice and inspire even deeper exploration into this interesting field.

1. Measurement and Cerebral Blood Flow and Oxygen Consumption, and the Regulation of Cerebral Circulation

Experimental Studies of Cerebral Blood Flow

Measurements of cerebral blood blow (CBF) is based on the use of *freely diffusible indicators*, which reach the brain tissue by the arterial system and give rise to a fast and complete equilibration in concentration between blood and tissue. The principle of calculation of CBF is based on the measurement of *mean transit time*. The technique has been developed during the last 40 years and was originally introduced by Kety and Schmidt (1945) with the use of *nitrous oxide* as tracer. Later, the technique was elaborated to include diffusible indicators including *Krypton-85* and *Xenon-133*. The Kety-Schmidt and the intraarterial 133-Xenon methods have been used in several animal experiments. By surgical removal of soft tissue over the calvaria and 133-Xenon injection in the lingual artery, it is possible to avoid extracerebral contamination in monkeys and rats (Harper and Jennett 1968, Hertz *et al.* 1977).

The *Hydrogen clearance* technique introduced by Aukland *et al.* (1964) has some advantages in experimental studies, because of its ability to obtain multiple flow measurements over long periods of time (up to 10 hours), the ability to measure flow in a small tissue volume, and the stability of the partition coefficient, especially in damaged tissue. The measurement of rCBF is based on a linear function between electrode current and tissue hydrogen concentration, provided that the thickness of the diffusion layer is constant. Comparisons with other methodologies (venous outflow, 133-Xe, radioactive microspheres and 14-C antipyrine) have shown fairly good correlations (Rowan *et al.* 1975, LaMorgese *et al.* 1975, Heiss and Traupe 1981). On the other hand, the technique is traumatic, giving rise to tissue injury due to electrode implantation and resulting in loss of autoregulation, hyperaemia and oedema (Tuor and Farrar 1984). Moreover, zero baseline stability can be difficult to obtain; arterial recirculation of hydrogen and intercompartmental diffusion might be causes of error as well (for review see Farrar 1987).

Microsphere cerebral blood flow determination was introduced by Roth *et al.* (1970). A number of conditions must be fulfilled for the accurate reflection of CBF by microspheres. The microspheres must be well-mixed at the injection site, the distribution in the blood stream must be proportionally to the actual blood flow, the microspheres must be trapped completely on first passage and not disturb the regional or general circulation, and they must be stably lodged until counted. Currently, 10 isotopes for labelling microspheres are available. Each isotop is characterized by specific gamma energy peak, thereby allowing detection by differential spectrometry. Organ blood flow in relative terms is calculated as the percentage of spheres in the area of interest in relation to the total number of spheres injected. In absolute terms, blood flow can be calculated by measuring cardiac output by a separate technique (Nuetze *et al.* 1968, Mendell and Hollenberg 1971), or by determining a reference organ blood flow by withdrawing blood at a constant rate during microsphere injection (Makowski *et al.* 1968, Domenech *et al.* 1969). Under normal physiological conditions a good correlations between CBF determined by microsphere technique and that obtained with 133-Xenon, iodoantipyrine, and hydrogen clearance have been found (Fan *et al.* 1979, Horton *et al.* 1980, Marcus *et al.* 1981, Heiss and Traupe 1981). Under pathologic conditions a good correlation was found with the 133-Xenon technique at flow rates below 120 ml/100 g/min, but not above this flow rate (Marcus *et al.* 1981). During middle cerebral artery occlusion, flow determined by the microsphere method consistently showed higher flow values than those obtained with the hydrogen clearance method (Heiss and Traupe 1981). In anaesthetized and awake animals the microsphere technique works excellently. Both regional and whole brain blood flow are easily obtained, and the calculation of flow is not dependent on a diffusion coefficient. Shunting of microspheres might occur, especially under pathological conditions and during anaesthesia. It must be stressed that the use of microspheres in various pathologic models has not been fully evaluated.

Venous outflow from the portion of the brain drained by the sagital and straight sinuses can be drained from the torcula. If the lateral sinuses and the occipital emissary veins are occluded, the venous outflow represents CBF. From the confluence of the sinuses the blood passes through a transducer probe or an electromagnetic flowmeter. This method was intraduced by Rapela and Green (1964) and was further elaborated in the dog by Michenfelder *et al.* (1968) and Michenfelder and Theye (1968), who found a good correlation between flow determined by the venous outflow technique and that obtained by the 133-Xenon method. In dogs, the venous outflow obtained represents drainage from 43% of the brain. In rats continuous venous outflow can be measured after cannulation of the retroglenoid vein (Meldrum and Nilsson 1976, Nilsson and Siesjö 1983). This method has been investigated by Morii *et al.* (1986 a), who found minimal extracerebral contamination and a good correlation to flow obtained by the microsphere technique.

The *autoradiographic* method of measurement of local CBF is based on Kety's study of the kinetics of inert gas exchange. The method was introduced by Landau *et al.* (1955) and Freygang and Sokoloff (1958) using 131-J trifluoroiodomethane as tracer. In 1969 Reivich and coworkers introduced antipyrine-14C, whilst Sakurada *et al.* (1978) introduced iodo (14C) antipyrine, which has a greater blood-brain barrier permeability compared with antipyrine-14C. The calculation of local CBF is determined from the concentration of the tracer in the region of interest. The values of the arterial concentration of the tracer apply under the assumptions that the tracer is biologically inert, the tracer in the effluent veins of the tissue is in equilibrium with that of the brain, the CBF remains in a steady state during the period of measurement, and the value of the partition coefficient of the tracer can be determined. It has been shown that autoradiographic strategy tends to underestimate flow, especially at increased flow rates (Eklöf *et al.* 1974) and under ischaemic conditions. Tomita and Gotoh (1981) have stressed that the method is inaccurate because of incomplete tracer mixing in ischaemic tissue and altered tracer permeability. However, with the use of very short tracer infusion periods, local CBF can be measured by a modification of the indicator fractionation technique elaborated by Goldman and Sapirstein (1973), and this technique has been used autoradiographically to estimate local CBF even in small animals (Pulsinelli *et al.* 1982).

After the development of the 2-deoxyglucose method for measurement of regional cerebral glucose utilization (rCMRgl) (Sokoloff *et al.* 1977), matched animal series for studies of rCBF and rCMRgl have been possible. This double-tracer autoradiographic strategy has recently been developed (Lear *et al.* 1981, Ginsberg *et al.* 1986), and Lear *et al.* (1984) have developed a generalized mathematic approach that allows two or more radionuclides with different half-lives to be used simultaneously to measure multiple aspects of cerebral circulation and metabolism. In the method of Sokoloff for rCMRgl determination, local 2-deoxy glucose accumulation is the basis for estimation of local glucose phosphorylation. However, only in normal tissue can the rate of 2-deoxyglucose accumulation be assumed to be a known fraction of glucose consumption. If the relationship between plasma and brain glucose has changed, a correction ratio of 2-deoxcyglucose accumulation to glucose consumption (the lumped constant) might be determined separately by 3-0 methylglucose autoradiography. Methods for this triple-tracer technique has been developed by Gjedde and Diemer (1983).

Measurement of Cerebral Blood Flow (Human Studies)

Measurement of CBF in humans was introduced by Kety and Schmidt with nitrous oxide as freely diffusible indicator (Kety and Schmidt 1945, Kety and Schmidt 1948). The method as originally described by Kety, presumes catheterization of the internal jugular vein, either by insertion of a catheter from the lateral or the anterior approach of the neck with the tip of the catheter directed cranially and placed at the base of the skull. Furthermore, a catheter in a peripheral artery is necessary. During a 15–30 min period of inhalation of the diffusible tracer, or during exhalation of the tracer after establishment of equilibrium between blood and brain tissue, samples of arterial and jugular venous blood are withdrawn at fixed intervals and the concentration of tracer determined. By the use of the height-over-area formula, CBF is calculated as ml/100 g/min as a global estimate of blood flow. By simultaneous determination of arterial and venous blood oxygen content, the arterio-venous oxygen content difference ($AVDO_2$) is calculated in vol%, and by simple multiplication the cerebral metabolic rate of oxygen

($CMRO_2$), also expressed as ml/100 g/min, is calculated.

The method has been criticezed because of contamination by extracranial venous blood (Lassen and Lane 1961) and central venous blood (Steinbach *et al.* 1976, Murray *et al.* 1978). Naturally, the method does not give information on the regional distribution of CBF. The method is time consuming and requires a contant CBF and cerebral metabolism for at least 15 min. Moreover, the results of CBF calculation are not immediatly available. On the other hand, the method gives a reliable estimate of global CBF and metabolism. Peroperatively, during intracranial surgery, the application of the method is fairly easy and the operative procedure is not disturbed. Sapirstein and Ogden (1956) have argued that erroneous results for CBF are obtained with this method, because the arterial and venous desaturation or saturation curves never reach each other during the period of investigation. This error results is an overestimation of CBF.

The Kety Schmidt technique has been developed into an intraarterial residue technique using Krypton-85 but now usually *Xenon-133* as tracer (Lassen and Ingvar 1963, Høedt-Rasmussen *et al.* 1966). This modification presumes catherization of the internal carotid artery. Owing to neurological complications thought to be due to microembolisation or spasms of the cerebral arteries, the method has however been critizised (Ingvar and Lassen 1973). Using meticulous technique and heparinized catheters the neurological complications have been found to be negligible (Lassen 1986). The principles of calculation are identical with the original Kety and Schmidt technique. With the use of externally placed scintillation detectors the technique has been modified to include 254 regional detectors (Sveinsdottir *et al.* 1977). Together with compartmental analysis, the method gives information on flow in grey and white matter, and the weights of these substances (Høedt-Rasmussen 1967). The method has been applied in clinical studies of cerebral diseases, including cerebral tumours, stroke, aneurysm, and head injury. Accordingly, important information concerning the regulation of CBF in relation to brain function and metabolic activity has been obtained (Ingvar and Lassen 1975). The normal value of global CBF based on 10 min saturation and measured in normothermic, normocapnic adult is 50 ml/100 g/min (Lassen and Munck 1955, Lassen 1959).

By analysis of the first two minutes of the semilogarithmically displaced clearance curve, Olesen *et al.* (1971) stated that this *initial slope* correlates fairly well with the values obtained by the principles of mean transit time. The initial slope method has been used extensively because it only requires 2 min data collection. In studies of head injury (HI) an initial fast component referred to as tissue peak has been described (Kasoff *et al.* 1972, Enevoldsen *et al.* 1976, Cold *et al.* 1977 b). Tissue peaks are observed in regions with cerebral contusion and subdural haematoma, and are thought to develop secondarily to tissue ischaemia. It must be emphazised that compared with compartmental analysis, the initial slope index underestimates grey matter flow by 20–30% (Lassen and Christensen 1976), and overestimates flow based on mean CBF. Consequently, the method does not give true values for $CMRO_2$. The normal value of CBF-initial in normothermic, normocapnic man is 65 ml/100 g/min (Olesen *et al.* 1971).

Inhalation and *intravenous* injection of 133 Xe have to some extent displaced the intraarterial approach. These two methods have been developed by Mallett and Veall (1963) and Agnoli *et al.* (1969) respectively and have been elaborated by others (Obrist *et al.* 1967, Austin *et al.* 1972, Risberg *et al.* 1975, Wyper and Brooke 1977). In comparison with the intra-arterial approach both methods are atraumatic; consequently, these methods have gained some popularity. Via externally placed scintillation detectors, often 10–32, regional CBF (rCBF) can be calculated. The calculation is complicated and cannot be done by visual analysis of the clearance curves. Corrections for flow in extracranial regions and remaining activity are necessary. However, the method has been accepted in clinical studies, also when repeated studies are performed. It has been stated that flows determined by compartmental analysis are less stable than non-compartmental indices, especially when grey matter flow is less than 50 ml (Obrist and Wilkonson 1985). Accordingly, Risberg *et al.* (1975) have elaborated an initial slope index (ISI index) based on the slope between 2 and 3 min. of the recirculation-corrected curve. With the inhalation technique the spatial resolution is fairly poor, the "cross-talk" from the uppersite hemisphere being a constant phenomenon (Wyper and Brooke 1977), and it has been argued that reliable results in clinical studies of focal lesions of ischaemic or flow deprivation are not possible (Halsey 1981). Thus, it has been argued that regions with low perfusion will be "looked through" due to higher radiation from adjacent tissue areas (Donley *et al.* 1975, Risberg 1980, Ingvar and Lassen 1982). However, other studies including evaluation of the sensitivity of the inhalation technique to ischaemic and probe-by-probe tests by use of ISI indices have shown that the inhalation technique is acceptable as regards ability to detect regions with ischaemic flow (Ewing et al. 1981, Ewing et al. 1983).

Positron emission tomography (PET) is based on positron emitting radionuclides, which after travelling a few millimeters in the tissue, interact with an electron, resulting in annihilition radiation with two gamma photons travelling in an uppersite direction. Pairs of externally placed detectors are positioned on either side of the gamma source and connected by electronic coincidence circuits, and only recording signals when two photons arrive within a very short time interval. In this way only photons arriving from positron annihilition are recorded (Phleps et al. 1975). A PET device consists of a large number of detector pairs. The information obtained from these detectors is combined to produce a two dimentional reconstruction of the regional radioactivity within the field (Ter-Pogossian 1977). PET permits *in vivo* measurements of rCBF, regional blood volume, regional oxygen utilization, and regional values of lambda (blood brain partition coefficient) (Kanno and Lassen 1979, Phleps et al. 1982, Huang et al. 1982, Raichle 1983). rCBF is measured during inhalation of 15-0-labelled CO_2 which is transferred to 15-0-labelled H_2O in the pulmonary vasculature. After 10 min of inhalation a steady state is achieved where the amount of tracer entering the brain equals the amount of leaving by washout and radioactive decay. Another modification is based on iv bolus injection of tracer (PET-autoradiographic approach). In comparison with the inhalation-steady-state method a linear correlation between radioactivity and rCBF is obtained, and no amplification error is seen at higher flows. Only a few PET scanning studies in patients with acute HI have been published (Alavi et al. 1989, Tenjin et al. 1989).

In comparison with PET, unpaired detectors are used during *Single-photon emission computed tomography (SPECT)*. SPECT has been designed for regional studies using 133-Xenon inhalation (Stokely et al. 1980). After 133 Xe inhalation for 1–2 min, a series of tomograms are produced (Lassen et al. 1981). This method yields a resolution of 1.5 to 1.7 cm (Holm et al. 1985). With the introduction of iodine-, thallium- and technetium-labelled tracers SPECT has been shown to play an important role in the diagnosis and management of cerebral diseases (For review, see Holman and Hill 1987). Only one study of SPECT in the acute phase of head injury has been published. Compared with CT scanning, SPECT is stated to have the following advantages. By reflecting changes in cerebral perfusion SPECT is more sensitive than CT scanning in demonstrating cerebral lesions, and SPECT demonstrates the lesions at an earlier stage than does CT scanning (Abdel-Dayem et al. 1987).

Using *Enhancement of CT scans* during inhalation of 30–50% stable Xenon, an atraumatic tomography approach to rCBF mapping is possible (Winkler et al. 1977, Drayer et al. 1978). This method has been developed by several groups and has been used in experimental studies of focal ischaemia (Yonas et al. 1988), in clinical studies of head injury (Yonas et al. 1984 c, Wozney et al. 1985, Harrington et al. 1986, Darby et al. 1988), cerebral aneurysms (Yonas et al. 1984 b), cerebral infracts (Drayer et al. 1980, Yonas et al. 1984 a) and multiple-vessel occlusion (Yonas et al. 1985). With the CT scanning equipment available rCBF mapping 5 min after stable Xenon inhalation is possible, each voxel measuring $1 \times 1 \times 5$ mm. The method has the interesting advantage that it is possible to calculate both the regional distribution of rCBF and that of the partition coefficient (lambda) (Meyer et al. 1980, Dhawan et al. 1984). Thus, preliminary communications indicate that the method can provide important information on the occurrence of cerebral oedema and infarction (Segawa et al. 1983, Yonas et al. 1984 b). On the other hand, Xenon, an anaestetic gas, induces an increase in CBF (Junck et al. 1985, Gur et al. 1985, Obrist et al. 1985), an increase in the CO_2 reactivity and an impairment of the cerebral autoregulation (Hartmann et al. 1987). Furthermore, stable Xenon inhalation gives rise to respiratory irregularities with bradypnoe and bronchospasms and changes in the state of consciousness with risks of vomiting and aspiration (Yonas et al. 1987).

Human Studies of Cerebral Metabolic Rate of Oxygen (CMRO$_2$)

With the nitrous oxide method for the quantitative determination of CMRO$_2$, Kety and Schmidt (1948) found that cerebral oxygen uptake averaged 3.5 ml 02/100 g/min. In a later study by Lassen et al. (1960), CMRO$_2$ averaged 3.0 ml. In his review from 1959, Lassen stated that no sexual or racial differences in CMRO$_2$ excisted. However, several studies indicate a small but significant decrease in CMRO$_2$ in elderly awake subjects, and a higher level of CMRO$_2$ in children (Kennedy and Sokoloff 1957). Sleep does not influence CMRO$_2$ significantly (Mangold et al. 1955).

A coupling between CBF and metabolism has repeatedly been shown. Thus Olesen (1971) found that CBF increased significantly in the part of the cerebral cortex representing the hand during maximal physical efforts with the contralateral arm, and listening to speech activates cortical areas in both hemispheres (Larsen et al. 1977). Similarly, epileptic seizure activity is associated with an increased cerebral oxygen uptake (Brodersen et al. 1973), cerebral energy depletion (Collins et al. 1970) and an increase in the concentration of lactate in brain tissue (Beresford et al. 1969, Bolwig and Quistorff 1973).

In patients suffering from organic dementia $CMRO_2$ is reduced (Lassen 1959), and in demented patients with communicating hydrocephalus an increase in CBF and $CMRO_2$ has been observed after implantation of a shunt (Greitz et al. 1969, Malmlund et al. 1972). In patients with severe head injury, a significant decrease in $CMRO_2$ parallels the depression consciousness (Obrist et al. 1984), and $CMRO_2$ as low as 0.4 ml $02/100$ g/ min has been found in the acute phase after head injury in patients who recover (Cold et al. 1978). The dynamic changes in $CMRO_2$ in patients with HI are discussed on page 37.

Hypothermia

Hypothermia decreases $CMRO_2$ and CBF proportionally (Rosomoff and Holaday 1954). In the temperature range 25–37 °C linear correlation between log $CMRO_2$ and body temperature has been found. Based on animal studies the constancy of the ratio delta log $CMRO_2$/delta temp Celsius ranges from 0.038 to 0.076, indicating a percentage change in $CMRO_2$ ranging from 9.1 to 19.1% per degree change in temperature (Feruglio et al. 1954, Adams et al. 1957, Bering 1961, Cohen et al. 1964, Michenfelder and Theye 1968, Tabaddor et al. 1972, Bering 1974, Astrup et al. 1981 b).

Hypothermia exerts a protective effect on the brain in anoxia (Boyd and Connolly 1961) in hypoxic hypoxia (Carlsson et al. 1976), and in experimental head injury (Bouzarth et al. 1967), and it has been suggested that even a small decrease in temperatur of 1–3 degrees Celcius might have a protective effect in experimental brain ischaemia (Marshall et al. 1956, Kramer et al. 1968, Michenfelder and Theye 1970, Kopf et al. 1975, Berntman et al. 1981, Young et al. 1983, Busto et al. 1987). On the other hand, even small increments in temperature in ischaemic brain tissue seem to accentuate histopathologic changes (Busto et al. 1987). A barbiturate in combination with hypothermia might

effectively control intracranial hypertension in some patients with severe head injury (Shapiro et al. 1974). Likewise, the combination of barbiturate and hypothermia might accentuate the protective effect in brain ischaemia (Hägerdal et al. 1978, Lafferty et al. 1978, Nordström and Rehncrona 1978). However, animal experiments during anoxia indicate that although barbiturate and hypothermia might have potentially similar effects on $CMRO_2$, dissimilar mechnisms might be involved, because the rate of ATP depletion and lactate accumulation is considerably lower during hypothermia (Michenfelder and Theye 1970).

Among the consequence of cerebral ischaemia known to be temperature sensitive are intracellular hydrogen ion homeostasis, calcium influx into neuronal structures, degradation of membrane lipids, and permeability of the blood-brain barrier (BBB) (Norwood and Norwood 1982, Krantis 1983, Rossi and Britt 1984, Lantos et al. 1986). Furthermore, hypothermia is thought to inhibit various neurotransmitters (Vanhoutte et al. 1981, Boels et al. 1985, Haikala et al. 1986), whereas other studies suggest an accelerated metabolic turnover (Okuda et al. 1986). It has been suggested that the protective effect of hypothermia is based on two mechanisms. 1) since hypothermia shifts the oxygen haemoglobin-dissociation curve towards the left, and prevents or minimizes a rightward shift due to acidosis, it maintains a high oxygen content in arterial blood. 2) by reducing $CMRO_2$ and consequently cellular energy requirements, hypothermia exerts a protective effect on the cellular level (Carlsson et al. 1976 b). However, experiments during hypoxia in rats indicate that the increased oxygen content in arterial blood produced by hypothermia is not a major determinant in hypothermic protection (Keykhah et al. 1980), and generally the protective effect is supposed to be caused by the decrease in $CMRO_2$.

Based on experimental studies the use of hypothermia has been proposed in the intensive care of patients suffering from brain ischaemia (Cohen 1981), and hypothermia has been suggested in the treatment of cerebral ischaemia in humans (Connolly et al. 1962), during extracorporal circulation (duCailar et al. 1964), after circulatory arrest (White 1972), during neurosurgical operations (McKissock et al. 1960, Terry et al. 1962, Uihlein et al. 1962, Daw et al. 1964, Deligné and David 1966, Uihlein et al. 1966, White et al. 1967) and in the treatment of severe head injury (Sedzimir 1959, Shapiro et al. 1974). During the last decade, however, the once common use of hypothermia has been abandoned, and later experimental studies of hypothermia in the treatment of acute stroke have not shown any beneficial effect; on the contrary, a detrimental effect was observed in primates and cats (Michenfelder and Milde 1977, Steen *et al*. 1979). It must be stressed that clinically controlled studies of the effect of hypothermia on clinical outcome in patients with neuroaxial trauma have never been published.

CO_2 Reactivity and Adaptation to Prolonged Hyperventilation

The cerebrovascular resistance (CVR) is primarily regulated by changes in pH in the extracellular fluid surrounding cerebral vessels (Betz and Heuser 1967, Cotev and Severinghaus 1969, Plum and Siesjö 1975). Thus, an increase in pH constricts and a decrease dilates cerebral resistance vessels. A decrease in CVR results in an increase in CBF and CBV, and an increase in CVR gives rise to an increase in flow and blood volume. As CO_2, in comparison with bicarbonate, is freely diffusible between arterial blood and brain tissue, changes in ventilation and consequently $PaCO_2$ are of utmost importance in the regulation of CBF. Preliminary studies in rats suggest that endogenously released adenosine is involved in the increase in CBF during hypercapnia (Phillis and De Long 1987). Although indomethacin, a cyclo-oxygenase inhibitor, reduces the increase in CBF associated with hypercapnia (Dahlgren *et al*. 1981), recent studies do not indicate that prostaglandins are involved in the regulation of CBF during hypercapnia (Wei *et al*. 1980).

The absolute CO_2 reactivity is defined as delta CBF/ delta $PaCO_2$ mm Hg. In the range of 4–6 kPa of $PaCO_2$ absolute CO_2 reactivity is 1–2 ml/mm Hg (Reivich 1964, McHenry *et al*. 1965, Raichle and Plum 1972). The relative CO_2 reactivity is defined as %change CBF/ delta $PaCO_2$ mm Hg or delta ln CBF/delta $PaCO_2$ mm Hg. In normal subjects it averages 4%/mm Hg at $PaCO_2$ ranging from 5–6 kPa (Olesen *et al*. 1971). During hypocapnia, $PaCO_2$ ranging between 3 and 5 kPa, the relative CO_2 reactivity is 2%, and during hypercapnia ($PaCO_2 > 6$ kPa) it is 6% (Tominaga *et al*. 1976). Comparative studies of regional CO_2 reactivity suggest that it is highest in the cerebrum, less pronounced in the cerebellum ($\times$ 0.6), and lowest in the spinal cord ($\times$ 0.5) (Sato *et al*. 1984).

Animal Experiments

Rosomoff (1963) studied changes in CBV and CSF volume after 30 min of hypocapnia to $PaCO_2$ 20 mm Hg in dogs and found a fall in CBV and an compensatory increase in CSF volume. The changes in CBV effected by hypocapnia and hypercapnia are correlated to changes in CBF. In normal brains CBV changes by about 0.04 ml/100 g/mm Hg $PaCO_2$. For the total brain, this means an increase of 11 ml in CBV when $PaCO_2$ is increased from 4.0 to 6.7 kPa (Grubb *et al*. 1974). During active hyperventilation cerebral venous drainage is not impaired, but during massive passive hyperventilation in dogs a high positive airway pressure impedes cerebral venous drainage, increases cerebral venous pressure, and consequently increases ICP (Kitahata *et al*. 1971).

In cats subjected to hypocapnia of 2.7 kPa, an increase in the concentration of lactate in cerebral tissue and CSF has been observed. It has been argued that this increase does not indicate cerebral ischaemia, as ATP and phosphocreatine are not depleted (Granholm and Siesjö 1969); however, following prolonged hyperventilation to $PaCO_2$ ranging from 1.3 to 1.6 kPa, a depletion of ATP and phophocreatine has been observed in cats and dogs (Granholm and Siesjö 1969, Michenfelder *et al*. 1970), and in addition a decrease in $CMRO_2$ occurs (Grote *et al*. 1981). Studies of brain tissue pH during respiratory alkalosis in dogs (Hilberman *et al*. 1984), and studies of cerebral tissue oxygen tension (Kennealy *et al*. 1980) support the view that extreme hypocapnia might provoke cerebral ischaemia. In a recent study by Samra *et al*. (1989) local glucose utilization was increased heterogenously throughout the brain in rats subjected to hypocpnia ($PaCO_2$ 25 mm Hg), with a significant increase in the lateral and ventral thalamus, the inferior colliculus, lateral habenulla, medial geniculate body, and auditory cortex.

Experimental studies of ischaemia have shown a decrease in focal pial blood pressure during hypercapnia (Brawley *et al*. 1967). This phenomenon is caused by a redistribution of blood flow from regions with a relatively high ICP and low CO_2 reactivity to regions with a high CO_2 reactivity and relatively low tissue pressure, referred to as a *steal phenomenon*. In contrast, hypocapnia might redistribute blood flow from regions with low tissue pressure and high CO_2 reactivity to regions with high pressures and relatively low CO_2 reactivity, *inverse steal phenomenon*. The occurence of the inverse steal phenomenon is of considerably interest and has focused attention on hypocapnia as a therapeutic tool in experimental brain ischaemia. Preliminary studies in dogs and cats suggest that the size of an infarct can be reduced if hypocapnia is applied prior to the insult (Soloway *et al*. 1968, Battistini *et al*. 1969).

However, later experimental studies have not corroborated this finding (Soloway et al. 1971).

In studies in dogs and goats, adaptation to prolonged continuous hypocapnia occurs within 2–3 hours (Raichle et al. 1970, Albrecht et al. 1987). Dogs subjected to normoxic hypercapnia initially show a dramatic increase in CBF accompanied by a decrease in CVR. The increase in CBF persists for three hours and is followed by decrease in CBF and an increase in CVR. Regional variations occur. Areas with the highest initial CBF show a greater rate of decay in flow over time. CSF-pH, initially more acid during hypercapnia, increases with time, accompanied by an increase in CSF-bicarbonate. Recently, Hansen et al. (1986) in studies in piglets subjected to prolonged hypocapnia found an increase in CBF after 30 minutes. During sustained hypocapnia in dogs, total and regional CBF correlates well with CSF-pH and a shift in the cerebrovascular sensitivity to CSF-pH has been found (Warner et al. 1987), Studies of pial arteriolar diameter by the cranial window technique in rabbits subjected to prolonged contiuous hypocapnia indicate that hypocapnia is only effective in reducing pial vessel diameter for less than 24 hours. The authors argue that hyperventilation in clinical practice should only be used when ICP is elevated. If used preventively, its effect may have dissipated by the time ICP starts to rise (Muizelaar et al. 1988).

Prolonged hypocapnia decreases the rate of formation of CSF (VF); however, after an initial decrease at 30 and 60 min, VF returns to prehypocapnic values (Martins et al. 1976, Hochwald et al. 1976, Artru and Hornbein 1987). In dogs with an intracranial mass-expanding lesion studies indicate that prolonged hypocapnia initially gives rise to a decrease in CBV. However, the CSF-pressure lowering effect is sustained by further reduction in CSF volume, despite reexpanding of the CBV. In the same model brain water content did not contribute to changes in CSF pressure and volume (Artru 1987).

Human Studies

In awake, unsedated patients, active hyperventilation to $PaCO_2$ 2.7 kPa induces changes in the EEG compatible with cerebral hypoxia (Morgan and Ward 1970), and these changes disappeare when hyperbaric oxygenation is provided (Reivich et al. 1966). Accordingly, during prolonged hyperventilation in comatose patients an improvement of consciousness has been observed when $PaCO_2$ became normal (Froman 1968).

On the basis of theoretical considerations proposed by Sørensen (1978), hypoxaemia to PaO_2 4.0 kPa, a level at which consciousness is altered and eventually lost, equals hyperventilation to $PaCO_2$ 3.5 kPa at a $CMRO_2$ value of 3.0 ml/100 g/min, as regards the decrease in jugular venous oxygen tension which under these conditions should average 3.0 kPa. At this level signs of cerebral hypoxaemia are evident (Allen and Morris 1962, Wollman and Orkim 1968). The threshold at which deterioration in consciusness occurs and EEG signs compatible with cerebral hypoxaemia are observed is a jugular venous oxygen tension of about 2.5–3.0 kPa. These low tensions occur at PO_2 ranging between 3.5 to 4.0 kPa and at a $PaCO_2$ levels of 2.5–3.0 kPa. In humans subjected to extreme hypocapnia, a moderate increase in CMR-glucose has been found, indicating anaerobic cerebral metabolism (Alexander et al. 1968). This change occurs at CBF levels ranging from 10 to 20 ml/100 g/min, and at jugular venous oxygen tensions of 2.7 kPa (Gotoh et al. 1965).

The decrease in oxygen delivery capacity effected by hypocapnia is partly caused effected by the decrease in CBF (part 75%) and partly by a shift of the dissociation curve of oxyhaemoglobin (part 25%) (Cain 1963, Gotoh et al. 1965, Harp and Wollman 1973).

In several studies of HI the CO_2 reactivity has been shown to be low in the acute phase and that a low hemispheric CO_2 reactivity is correlated to a poor outcome (Fieschi et al. 1974, Overgaard and Tweed 1974, Cold et al. 1977 a, Messeter et al. 1986). In patients with intracranial hypertension, hypocapnia effectively reduces ICP and CBF. It is accepted that the fall in ICP is caused by vasoconstriction of cerebral vessels and a decrease in CBV. Although the decrease in CBV is small in comparison with the total brain volume, hypocapnia can be life-saving in patients with an expanding cerebral lesion, and acute hyperventilation is, therefore, an important tool in the management of acute intracranial hypertension (Lundberg et al. 1959, Slocum et al. 1961, Bozza et al. 1961).

In patients with apoplexy and cerebral tumours, hypercapnia might provoke a steal phenomenon by promoting a decrease in CBF in the focal region of incomplete or complete ischaemia (Palvölgyi 1969, Paulson 1970). On the other hand, in severe head injury, apoplexy, and brain tumours, an inverse steal phenomenon or Robin Hood phenomenon has been observed during hypocapnia (Palvölgyi 1969, Paulson 1970, Pistolese et al. 1972, Fieschi et al. 1974, Cold et al. 1977 b, Obrist et al. 1984, Darby et al. 1988). In Fig. 1. an example of inverse steal reaction is shown. The patient,

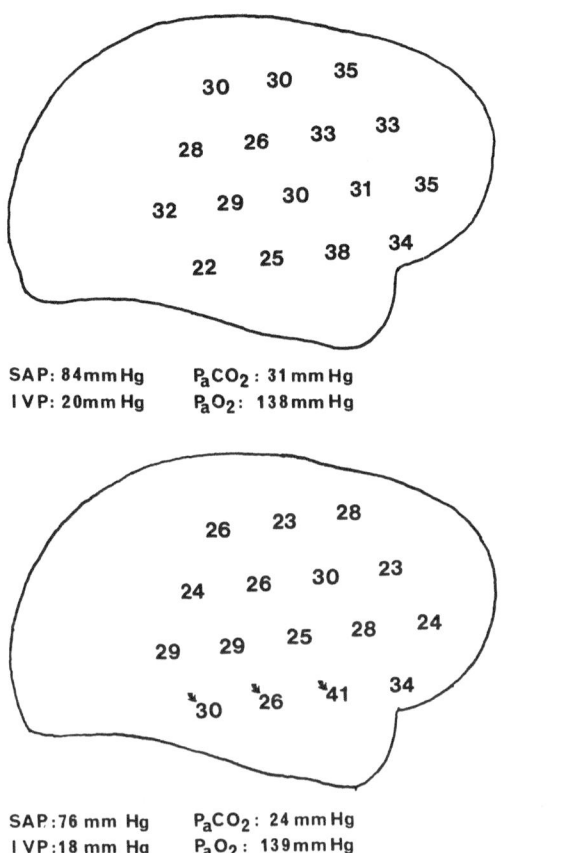

SAP: 84 mm Hg P_aCO_2: 31 mm Hg
IVP: 20 mm Hg P_aO_2: 138 mm Hg

SAP: 76 mm Hg P_aCO_2: 24 mm Hg
IVP: 18 mm Hg P_aO_2: 139 mm Hg

Fig. 1. rCBF in a 23-year-old unconscious man with contusion and laceration of the right temporal lobe. After reduction of $PaCO_2$ from 31 to 24 mm Hg, an inverse steal reaction was observed in the right temporal region, indicated by arrows (Acta Anaesthesiol Scand 1977: 21: 359–367, with permission)

a 23-year-old-male, was examined on the third day after a severe head injury, just before a craniotomy, which showd contusion of the right temporal lobe. After reduction of $PaCO_2$ from 34 to 24 mm Hg an inverse steal phenomenon was found in the temporal lobe, indicated by arrows.

Occurence of the inverse steal reaction is of considerable interest. Studies of CA during hypocapnia which show a normalization of the CA during hypocapnia in patients with apoplexia and cerebral tumours (Paulson et al. 1972) and studies of hypocapnia as a therapeutic tool in the control of intracranial hypertension have suggested prolonged artificial hyperventilation as a rational treatment in patients suffering from brain ischaemia. However, controlled studies of the effect of artificial hyperventilation in cerebral apoplexy have been discouraging, no improvement in recovery or outcome being demonstrated (Christensen 1976). In pa-

tients with severe traumatic head injury an uncontrolled study of artificial hyperventilation suggests improvement of outcome (Gordon 1979). In a preliminary controlled study, however, prophylactic artificial hyperventilation resulted in a poorer outcome at three and six months after the trauma (Ward et al. 1989). The effects of acute hypocapnia on CBF and ICP in patients with severe head injury will be discussed on page 40.

The adaptation to prolonged hypocapnia has been investigated in patients with apoplexy, The half-life of the adaptation mechanism of CSF-pH and CSF bicarbonate averages 6 hours, and adaptation is said to be complete within 24–30 hours (Christensen et al. 1974). Recently, studies using non-invasive doppler ultrasound technique and calculation of the instantaneous mean blood velocity during hypocapnia in normal subjects indicated that blood velocity showed adaptation within 10 min after induction of hypocapnia (Ellingsen et al. 1987). However, in patients with severe head injury an uncontrolled study did not reveal signs of CSF-pH adaption within a period of 6–24 hours, and it is suggested that the mechanism of adaptation may be impaired in severe generalized cerebral lesions with lactic-acidosis or ischaemia (Cold et al. 1977 a). Fig 2. indicates the effects of sustained hypocapnia on ventricular fluid lactate, pyruvate, lactate/pyruvate ratio, bicarbonate, and pH in six unconscious patients suffering from severe head injury. Arterial and ventricular fluid samples were taken 24 hours before and after induction of hypocapnia. As indicated, no signs of CSF-pH adaptation were seen.

The Use of Prolonged Continuous Hyperventilation. Pro et contra

Some arguments for and against the use of prolonged artificial hyperventilation have already been advanced. Other arguments against the use of prolonged artificial hyperventilation are as follows: 1) inhibition of oxygen delivery from oxyhaemoglobin to the tissue, because of a shift to the left of the saturation curve of haemoglobin (Bohr effect). Besides inpairment of the oxygen delivery to the brain this effect includes impairment of oxygen supply to the myocardium (Neill and Hattenhauer 1975). 2) decrease in cardiac output and blood pressure caused by a reduction in the central venous blood flow and diastolic filling. These effects are mediated via an increase in mean airway pressure causing an increase in CVP, cerebral venous pressure and a decrease in CPP. An increase in airway pressure and CVP, combined with a decrease in blood pressure

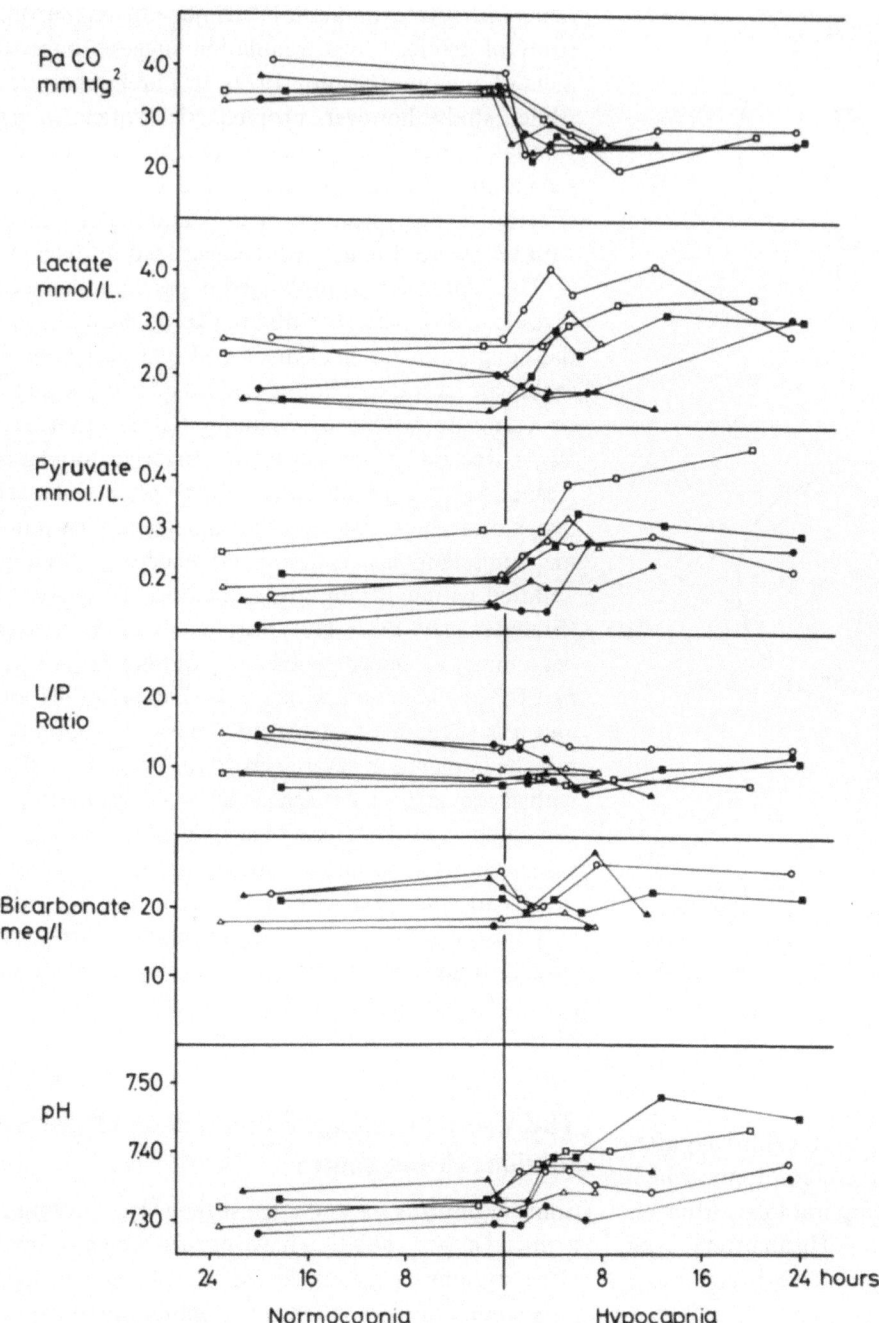

Fig. 2. $PaCO_2$ (mm Hg), ventricular fluid lactate (mmol/l), pyruvate (mmol/l), lactate pyruvate ratio, ventricular fluid bicarbonate (meq/l), and pH in six unconscious patients with head injury, 24 hours before and after sustained hypocapnia. The studies were performed 4–7 days after the acute injury. Before hyperventilation the ventricular fluid pH and lactate levels were normalized, and ICP were constantly below 20 mm Hg. The changes in $PaCO_2$ after induction of sustained hypocapnia are indicated. No sign of CSF-pH adaptation were observed within 8 hours in the six patients, and in 4 of 6 patients no sign of adaptation were observed within 24 hours (Acta Anaesthesiol Scand 1977: 222–231, with permission)

also induces a decrease in renal function and an associated retention of water and sodium. Lastly: The risk of barotrauma might be stressed as a function of increased airway pressure. The beneficial and detrimental effects of prolonged hyperventilation are summarized as follows:

Beneficial effect:

1: Decrease in ICP (control of ICP hypertension).
2: Respiratory alkalosis, neutralizing metabolic acidosis in extracellular tissue.

3: Normalization of cerebral autoregulation.

4: Inverse steal phenomenon (Robin Hood reaction).

5: Normalization of regional cerebral blood flow pattern (disappearance of tissue peaks). See page 40.

6: Prolonged hyperventilation reduces energy consuming and exhausting respiratory work.

7: Reduction of CSF formation.

8: The effect of prolonged hyperventilation on

CBV, CBF, and ICP is of prolonged duration (provided the adaptation mechanism is absent).

Detrimental effects:

1: Cerebral oligaemia in focal regions and watershed areas (See pages 78 and 81).
2: Decrease in diastolic filling and cardiac output.
3: Decrease in MABP and CPP.
4: Increase in CVP, and cerebral venous pressure.
5: Water and salt retension.
6: Inhibition of oxygen delivery to the tissues (Bohr effect).
7: Barotrauma.
8: The effect on CBF, CBV, and ICP is of short duration (Adaptation mechanism present).

Cerebral Autoregulation (CA)

Evidence of an intrinsic regulation of CBF designed to maintain a constant cerebral perfusion during changes in blood pressure was first presented by Fog (1934), and later confirmed by others (Forbes *et al.* 1937, Lassen 1959). Studies in cats with the cranial window technique for direct observation of the pial microcirculation indicate that adjustments in caliber within the autoregulatory range occur in vessels larger than 200 µm in diameter, whilst small arterioles less than 100 µm in diameter only dilate at pressures equal to or less than 90 mm Hg (Kontos *et al.* 1978). The *lower limit* of CA differs for different species. Below this level, CBF decreases and becomes pressure dependent; however, at pressures below the lower limit of CA maximal dilatation of the cerebral vessels does not occur. Thus, Häggendal and Johansson (1965) demonstrated in dogs that hypercapnia might increase CBF further. In cats subjected to mock CSF installation and examined by MR spectroscopy, lactate rises at an average CPP of 49 mm Hg and an increase in phosphocreatine has been observed at an average CPP of 29 mm Hg; however, considerable variation was found, in the CPP at which failure of brain energy metabolism occurred in cats, suggesting differences in the CA curves in this animal (Sutton *et al.* 1987). In dogs, the lower limit of CA is unchanged during hypocapnia (Artru *et al.* 1989). In dogs, an *upper limit* of CA has ben defined (Ekström-Jodal *et al.* 1972). This limit is also dependent on species and is increased in experimental renovascular hypertension in baboons (Strandgaard *et al.* 1975 b). In baboons CA regulation starts within 5–8 seconds of changes in MABP, and is largely completed within 20 seconds (Symon *et al.* 1973).

Above the upper limit, disruption of the *blood-brain barrier (BBB)* occurs and cerebral oedema and focal haemorrhage supervene (Johansson *et al.* 1970, Häggendal and Johansson 1972). Acute hypertension to this high level of blood pressure is associated with destructive endothelial lesions and abnormalities of the vascular smooth muscle in pial arterioles. These changes are inhibited by pretreatment with indomethacin or topical application on the brain surface of scavengers of free oxygen radicals (superoxide dismutase) (Kontos *et al.* 1981). Whilst hypertension above the upper limit of CA for about one hour is associated with long lasting distruption of the BBB and development of cerebral oedema (Ekström-Jodal *et al.* 1975), acute hypertension of short duration (<3 min) results in BBB disfunction that is reversible within 10 min (Johansson and Linder 1978). The disruption of the BBB primarily occurs in the venules and is presumably due to an increase in venous pressure (Auer 1978, Mayhan *et al.* 1986 a). Studies in cats suggest that the derangement of CA is heterogenous within the brain with relative resistance of the cerebellum (Sato *et al.* 1984). In contrast, acute hypertension in rats indicates that the hyperperfusion durcing acute increases in blood pressure is most prominent in the cerebellum, parietal grey matter, thalamus, striatum, and pons. These anatomical sites are recognized sites of hypertension hemorrhage in humans (Burke *et al.* 1987). Other studies in rats suggest that CA is more effective in the brain stem than in the cerebrum (MacKenzie *et al.* 1976, Suzuki *et al.* 1984, Baumbach and Heistad 1985, Murphy and Johanson 1985, Mayhan *et al.* 1986) and that the regional derangement of CA is influenced by stress, e.g. insulin-induced hypoglycemia and related to the occurence of neuropathological findings (Siesjö *et al.* 1983).

Studies in rats subjected to brain ischaemia by unilateral cerebral embolization by microsphere (Hardebo and Beley 1984) and studies in dogs with multifocal cerebral ischaemica induced by air embolism (Dutka *et al.* 1987) indicate that arterial hypertension during the ischaemic period increases the leakage of the BBB resulting in haemorrhage in the infarcted brain areas; Furthermore, acute hypertension induces delayed deterioration of brain function as evaluated by evoked response and worsens CBF. Other studies in gerbils subjected to carotid artery ligation have revealed that hypertension after glood flow restoration following ischaemia induces or accelerates BBB damage (Ito *et al.* 1980).

The CA has been studied extensively with the aid

of norepinephrine and angiotensin II. These drugs were assumed to have no action on the cerebral vessels (King *et al.* 1952, Agnoli *et al.* 1965, Greenfield and Tindall 1968, Olesen 1972) except when the MABP exceded the upper limit of CA and disruption of the BBB occurred (MacKenzie *et al.* 1976 ab, Edvinsson *et al.* 1978). Recent studies, however, indicate that these drugs have a direct constrictive effect on cerebral vessels (Sercombe *et al.* 1985, Speth and Harik 1985). Kuschinsky *et al.* (1982) have found a reduced cerebral glucose utilization and an increase in CBF during norepinephrine infusion. Furthermore, Reynier-Rebuffel *et al.* (1987) in studies in rabbits observed non-uniformity of CBF during norepinephrine or angiotensin II-induced hypertension suggesting that the cerebrovascular regulatory mechanisms in hypertension are more complex and cannot be explained completely by a simplet myogenic reaction as suggested by Symon *et al.* (1972). However, it is generally believed that myogenic reactions of smooth muscle in the arterial walls are responsible for CA regulation and in vitro studies of isolated arteries support this view (Vinall and Simeone 1981). Experimental studies suggest that adenosine plays an important role as a chemical mediator in CA regulation during induced hypotension (Winn *et al.* 1980, Winn *et al.* 1981). Released adenosine predominantly appears, however, to remain within the cerebral cells (McIlwain and Poll 1986) and contrary to the study by Winn *et al.* (1980) recent studies indicate that interstitial fluid adenosine concentration failed to increase during autoregulation studies in the rat (Van Wylen *et al.* 1987).

Activation of perivascular sympathetic nerves modifies the CA thresholds of lower and upper limits of CA by a shift towards higher levels (Rapela *et al.* 1967, Eklöf *et al.* 1971, Waltz *et al.* 1971). Other studies suggest that neurogenic activity influences the CA during stress (Salonga and Waltz 1973, Fitch *et al.* 1975, Bill *et al.* 1976, MacKenzie *et al.* 1979, Edvinsson *et al.* 1983). Moreover, studies in rats have shown that increased sympathetic activity during acute hypertension seems to blunt the vulnerability of the BBB (Johansson and Auer 1983).

The CA is easily abolished when vasodilatation of cerebral vessels is induced by papaverine (Johansson 1974), during hypercapnia and hypoxia (Häggendal and Johansson 1965).

In *Human studies* of CA the lower limit of CA lies between 50 and 80 mm Hg (Shenkin *et al.* 1950, McKrell *et al.* 1955, Kleinerman *et al.* 1958). Generally a level of 60 mm Hg is supposed to be the lower limit.

An upper limit of CA has been demonstrated in normotensive subjects (Strandgaard *et al.* 1975 a) and is thought to be about 150–170 mm Hg of MABP (Lassen 1974). In human CA regulation starts within 5–8 seconds and is completed within 20 seconds (Greenfield *et al.* 1984). The human equivalent to disruption of the BBB is acute hypertensive encephalopathy (Skinhøj and Strandgaard 1973). In chronic hypertension the lower and upper limits of CA are shifted to the right (Strandgaard *et al.* 1973); however, after 2–3 weeks of antihypertensive therapy a shift of the CA curve to the left has been observed (Strandgaard 1976). For review see Strandgaard and Paulson (1984).

Drugs with a vasodilatory capacity on cerebral vessels like papaverine and inhalation agents (halothane, enflurane and isoflurane) impair the CA. Impairment of CA has also been demonstrated during hyperthermia (Allen *et al.* 1986). On the other hand, normalization of $PaCO_2$ and hypnotic drug, which induce a suppression of cerebral oxygen uptake, will restore CA. Studies in humans with cerebral tumours or apoplectic insults have shown that induced hypocapnia restores CA which otherwise is abolished during normocapnia (Paulson *et al.* 1972).

In the acute phase of head injury, CA generally is impaired. This impairment might be focal, hemispheric or global. However, several studies of CA in the acute phase of HI indicate that CA might be unimpaired

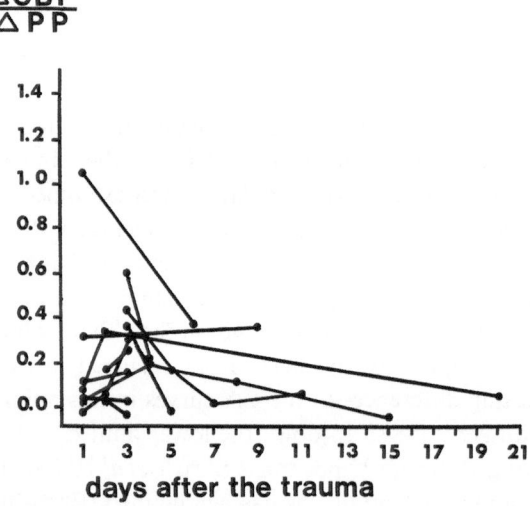

Fig. 3. Cerebral autoregulation (CA) indicated by the ratio: change CBF (initial slope)/change CPP mm Hg (perfusion pressure). CPP was defined as the difference between mean arterial pressure in ICP. The studies were performed repeatedly in 12 comatose patients with head injury. Changes in mean arterial blood pressure were provoked by angiotensin (Acta Anaesthesiol Scand 1978. 22: 270–280, with permission)

during the acute phase (first days after the trauma), abolished later on, and might normalize during recovery. (Fieschi *et al.* 1974, Enevoldsen and Jensen 1978 b, Cold and Jensen 1978). In Fig. 3, the slope of the relationship between CPP and CBF (delta CPP/delta CBF) is correlated to time after the acute head injury. 18 patients were studied, and the studies were performed repeatedly. As indicated, the values of delta CPP/delta CBF with one exception are relatively small during the first day after the trauma, but increase later (Cold and Jensen 1978). As CA in experimental studies is very easely abolished, the initial unimpaired CA may to be a false phenomenon. The presence of "false" CA has been debated and it is thought that CA test via angiotensin-induced blood pressure increase, might provoke an increase in cerebral tissue pressure due to extravasation of water and electrolytes, which second-arily obstructs the cerebral vessels, leaving the CPP unchanged. Clinical studies, however, do not support this view since angiotensin-induced hypertension in patients with "false" CA is not always followed by an increase in ICP. In contrast, subjects with hemispheric or global impairment of CA generally show an increase in ICP. Therefore, either other mechanisms are responsible for the "false" reaction of CA or with available techniques it is not possible to accurately detect tissue pressure changes. In this context, studies of bifrontal subarachnoid pressure in patients with head injury and an unilateral supratentorial mass lesion indicate that no differences in pressures are found in comparative ICPs. These results suggest that a localized tissue pressure increase does not occur (Yano *et al.* 1987).

2. Ischaemic Thresholds and Development of Ischaemia

In experiments with reversible middle cerebral artery occlusion (MCAO) in cats and monkeys, reversible paralysis occurs when CBF is reduced to 23 ml/100 g/min and irreversible paralysis supervenes when CBF is permanently reduced to about 17–18 ml (Heiss *et al.* 1976, Jones *et al.* 1981). In studies of gerbils subjected to acute cerebral ischaemia by carotid ligation, a CBF threshold of CBF at 20 ml/100 g/min has been defined; below this threshold MR spectroscopy shows metabolic failure with phosphocreatine and ATP decline, combined with lactate accumulation and a fall in intracellular pH indication that failure of synaptic function is a direct consequence of energy failure (Crockard *et al.* 1987). In experiments with primates subjected to MCAO, pH in the extracellular space remains stable until CBF levels of 20 ml/100 g/min; below this value pH fell rapidly (Harris *et al.* 1987) and an increase in brain water was observed (Symon *et al.* 1979). If CBF falls to about 10 ml/100 g/min, further cytotoxic oedema occurs as a result of sodium passage into the cells. Unless the cells have been severely damaged, the formation of oedema is reversible (Hossmann 1976). Studies in baboons indicate that if the duration of ischaemia is too long, reperfusion will increase cerebral oedema (Bell *et al.* 1985).

In animal experiments the ischaemic threshold of evoked potential suppression is reach at CBF values of 15–17 ml/100 g/min (Branston *et al.* 1974, Heiss *et al.* 1976, Astrup *et al.* 1977). However, a difference in vulnerability has been observed with a lower threshold in the basal ganglia and the medial lemniscus in the brain stem (Okada *et al.* 1983, Branston *et al.* 1984). Furthermore, studies of spontaneous firing of nerve cells in the cat suggest differences in flow thresholds ranging from 6 to 20 ml/100 g/min (Heiss and Rosner 1983). If CBF is further reduced, below the level of 6–8 ml, efflux of K + and depletion of ATP occur (Astrup *et al.* 1977, Branston *et al.* 1977, Astrup *et al.* 1979) and calcium homeostasis is exhausted (Harris *et al.* 1981). At these low values cerebral cellular integrity and viability are severely impaired and cellular death and infarction threaten (Symon and Brierley 1976, Astrup 1982). Us-

ing MR spectroscopy these findings are in accordance with studies in gerbils indicating a decrease in intracellular pH at CBF levels of 18–23 ml, a decrease in ATP at CBF levels of 12–14 ml followed by disappearance of ATP and phosphocreatine and a decrease in 23 Na signal (Naritomi *et al.* 1988). In small animals (cats, rats, and gerbils) with a higher oxygen consumption, the threshold of membrane failure has been found to be higher (Hossmann and Schuier 1980, Strong *et al.* 1983, Mies *et al.* 1984, Harris and Symon 1984).

In CBF levels ranging from 8 to 23 ml, synaptic transmission failure threshold has been partly or completely exceeded, but membrane failure has not occured. This stage, called an ischaemic penumbra (Symon 1980, Astrup *et al.* 1981 a), has gained considerable interest because therapeutic measures aiming at an increase in CBF might theoretically reestablish synaptic transmission and might reduce neurological deficits. Studies in cats subjected to electrical cortical stimulation have shown that even in the penumbra after MCAO, some degree of flow reactivity persists suggesting that the microcirculation in the penumbra is not maximally vasodilated (Strong *et al.* 1988). For review see Astrup (1982) and Lassen and Astrup (1987).

The duration of flow deprivation influences the ischaemic threshold; hence, a CBF reduction to 8 ml is only tolerated for about one hour (Morawetz *et al.* 1979, Marcoux *et al.* 1982). However, neuronal transmission can be reestablished at flows of 10–12 ml for less than 2–3 hours. With permanent MCAO neuronal necrosis occurs at CBF in the range of 15–17 ml, and flows ranging from 18 top 23 ml are tolerated for more than two weeks.

If the ischaemic threshold is exceeded, cellular necrosis takes place and an infarct developes. During complete cerebral ischaemia, a selective vulnerability has been observed, involving the CA 1 subfield of the hippocampus, the caudo-putamen, the layers III–V in the cerebral neocortex, the cerebellar Purkinje cells, and the lateral reticular nucleus of the thalamus (Ito *et al.* 1975, Ginsberg *et al.* 1979, Diemer and Siemkowicz

1981, Pulsinelli *et al.* 1982 a, Blomquist and Wieloch 1985). In animal studies of .focal ischaemia after MCAO, a demarcated area of ischaemia develops in the occluded territory of the artery (Brierley and Graham 1984). In the peri-infarct zone, a mantle with reduced CBF, increased glucose consumption has been observed (Nedergaard *et al.* 1986). Spreading depression has been detected in these low perfusion areas, and this phenomenon is supposed to play a role in the development of selective neuronal damage (Nedergaard 1988).

The role of cerebral acidosis in the development of cerebral ischaemia has been debated. Using H + selective electrodes it has been shown by Thorn and Heitmann (1954) that the brain becomes acidotic during ischaemia. Furthermore, studies with H + selective microelectrodes indicated that the acidosis occurs in the interstitial tissue (Siemkowicz and Hansen 1981). Other experimental studies indicate that the degree of intracellular and interstitial acidosis contributes to neurological damage (Myers and Yamaguchi 1977, Nordström *et al.* 1978, Siemkowicz and Hansen 1981, Rehncrona *et al.* 1981). Recent studies indicate that glia cells become more acidotic than other brain cells (Kraig and Chesler 1987) and that an ischaemic infarction occurs when pH is less than 5.30 for more than 20 min (Kraig *et al.* 1987).

The level of the plasma-glucose influences the development of ischaemia. As demonstrated by Myers and Yamaguchi (1977) non-fasting monkeys subjected to global ischaemia develop more brain damage than fasting animals, and hyperglycaemic rats made a poorer recovery than did normoglycaemic and hypoglycaemic rats after 10 min of complete ischaemia (Siemkowicz and Hansen 1978). These observations have been extended by Rehncrona *et al.* (1981) who demonstrated that hyperglycaemic rats subjected to complete, in contrast to incomplete ischaemia, showed better recovery, that tissue lactic-acid accumulation especially developed during incomplete ischaemia, and that an lactate concentration exceeding 20 μmol/g was associated with impaired recovery. Furthermore, evidence is appearing which suggest that recovery in hyperglycaemic animals subjected to incomplete ischaemia is poorer than it is with complete ischaemia, and if hyperglycaemia develops prior to the insult, tissue lactic-acidosis has a detrimental effect on recovery (Siesjö 1984). Accordingly, in rats subjected to MCAO, tripple-tracer autoradiography has demonstrated that the threshold of local CBF for cerebral tissue pH reduction was lower in normoglycaemic compared with hyperglycaemic animals, and in the same study the threshold of local CBF

for local glucose hypermetabolism was found to be 20 and 30 ml/100 g/min in normoglycaemic and hyperglycaemic rats respectively (Nagai *et al.* 1988). However, the mechanism of the detrimental effects of hyperglycaemia is still unclear. Thus, in contrast to lactic-acid accumulation MR studies in rats indicate that hypercarbia-induced intracellular acidosis is tolerated well (Litt *et al.* 1985). Furthermore, the effect of pre-existing hypoglycaemia on ischaemic brain damage is inconclusive (Diemer and Siemkowicz 1981, Pulsinelli *et al.* 1982), and in focal ischaemia of MCAO in rats (Nedergaard and Diemer 1987), cats (Zasslow *et al.* 1987) and in rats subjected to photochemically induced cortical lesions (Ginsberg *et al.* 1987), hyperglycaemia diminishes the selective neuronal injury in the peri-infarct zone and the infarct size. Recent studies in rats subjected to MCAO with or without common carotid artery occlusion suggest that infarcted regions having collateral circulation are vulnerable to the deleterious effects of hyperglycaemia, whereas regions with end-arterial vascular supply are more resistant (Prado *et al.* 1988). Other studies in rats suggest that blood flow alterations after MCAO are not influenced by plasma glucose utilization. In contrast, glucose utilization depends on the combination of plasma glucose concentration and blood flow (Nedergaard *et al.* 1988).

As mentioned under CO_2 reactivity studies of hyperventilation in man indicate that EEG changes compatible with cerebral anoxia are seen at $PaCO_2$ level of 2.7 kPa, and these changes are normalized by hyperbaric oxygen. Furthermore, a moderate increase in CMR-glucose, indicating anaerotic metabolism, has been observed at extreme hyperventilation with CBF levels ranging from 10 to 20 ml, which correspond to jugular venous oxygen tensions of 2.7 kPa (Alexander *et al.* 1968). It is therefore thought that the ischaemic threshold for synaptic transmission in humas is reach at CBF levels of 20 ml/100 g/min.

Anaesthetic agents seem to influence the ischaemic threshold. During endarterectomy in patients anaesthetized with less than 1% halothane, EEG changes with flattening and isoelectric recordings have been observed during clamping when CBF falls to levels ranging between 18–23 ml (Trøjaborg and Boysen 1973, Sharbrough *et al.* 1973). An identical threshold of CBF has been found during carotid ligation in patients operated on for cerebral aneurysms (Leech *et al.* 1974) During neurolept anaesthesia, however, a higher stump pressure and lower CBF have been observed (McKay *et al.* 1976), and during isoflurane anaesthesia the ischaemic EEG changes are only observed when CBF is reduced to values below 10 ml (Messick *et al.* 1987).

3. The Effect of Barbiturate and Mannitol on CBF and Metabolism

Up to this point two principles regarding control of intracranial hypertension (hypothermia and prolonged artificial hyperventilation) have been discussed. The use of $Ca++$ blocking agents and indomethacin will not be discussed in this review, because at present no clinical studies in patients with severe head injury are available. On the other hand, barbiturate and mannitol have been used in the control of ICP-hypertension for many years. In the following the effects of barbiturate and mannitol on CBF and metabolism are reviewed.

Barbiturate: Experimental Studies

In dogs and rats a dose-dependent decrease in CBF and $CMRO_2$, associated with an increase in CVR has been observed after bolus injection of thiopentone (Michenfelder 1974, Carlsson et al. 1975, Stullken et al. 1977), and phenobarbitone (Nilssona and Siesjö 1975). Through their action on cerebral metabolism and flow, CBV and ICP will decrease. The suppression of $CMRO_2$ by thiopentone injection reachs a plateau of 50% on control values when the EEG is isoelectric (Michenfelder 1974, Kassell et al. 1980, Steen et al. 1983). However, in dogs subjected to thiopentone combined with hypothermia to 30°C, $CMRO_2$ decreases to about 70% of control values (Lafferty et al. 1978). At even greater degrees of hypothermia (14–18°C), the EEG becomes isoelectric, and $CMRO_2$ is reduced to 7–14% of control. Under these circumstances $CMRO_2$ is unchanged by barbiturate (Steen et al. 1983). During continuous thiopentone infusion in dogs and goats, loss of consciousness occurs when CBF and $CMRO_2$ are reduced by 23–30%, and emergence from anaesthesia is observed while $CMRO_2$ still is depressed by about 20% (Stullken et al. 1977, Albrecht et al. 1977). In dogs pre-treatment with thiopentone before induction af anaesthesia is associated with acute tolerance of $CMRO_2$ in spite of higher concentrations of thiopentone in CSF and blood (Altenburg et al. 1969). On the other hand, chronic thiopentone infusion for 24 hours in dogs is followed by a sustained increase in $CMRO_2$ occurring

after 3 hours in spite of unchanged blood levels of thiopentone (Gronert et al. 1981). In the rat, the effects of thiopentone on CBF and $CMRO_2$ are similar in both young and old animals (Baughman et al. 1986). Following high doses of thiopentone in dogs and rats (until an isoelectric EEG and a decrease in $CMRO_2$ averaging 50% are seen) the concentrations of ATP, ADP and AMP are unchanged. Simultaneously, the concentrations of glycogen and phosphocreatine are increased indicating an unchanged or increased energy charge level; thus, the cerebral metabolic effect of barbiturate is secondary to functional effects (Michenfelder 1974, Carlsson et al. 1975).

Studies of regional glucose utilization in rats indicate that barbiturate induces a universal metabolic depression, partly excluding nucleus interpeduncularis, nucleus habenula, and tractus habenulo-interpeduncularis (Herkenham 1981, Hodes et al. 1985). In dogs a shifting of the percentage contribution of flow to slow compartments (white matter) was observed after phenobarbitone infusion. It is suggested that this selective shunting of blood to white matter might explain the fall in ICP and the protection of deep white matter observed by many investigators (Laurent et al. 1982).

Studies in rats suggest an inhibition of cerebral phosphofructokinase activity by barbiturates, mobilization of endogenous substrates from existing carbohydrate and amino-acid pools, and reduction of bound hexokinase activity thereby inhibiting energy metabolism (Nilsson and Siesjö 1974, Chapman et al. 1978, Krieglstein et al. 1981). Other studies in rats indicate that intracellular pH is increased during barbiturate anaesthesia (Messeter et al. 1972, MacMillan and Siesjö 1973) and that the increase in intracellular pH is mediated by a decreased production of pyruvate and lactate through inhibition of phosphofructokinase (Carlsson et al. 1975, Chapman et al. 1978).

In baboons subjected to MCAO, an ultra-short acting barbiturate, methohexitone produces an elevation of CBF in cortical regions where flow is below evoked potential thereholds. This increase in flow most likely results from a reduction in flow in relatively well-perfused cortical regions as a consequence of a reduction in metabolic rate and vasoconstriction by the barbiturate, thereby diverting blood flow into relatively ischaemic regions (an *inverse steal*) (Branston *et al.* 1979). In cats subjected to MCAO, pentobarbitone infusion will increase cerebral oxygen availability in poorly perfused cerebral cortex (Feustel *et al.* 1981).

During high dose thiopentone infusion in dogs until burst suppression level in the EEG, a profound degree of vasoconstriction, equivalent to that produced by hypocapnia with $PaCO_2$ 20 mm Hg occurs. Under these circumstances CO_2 reactivity is fairly low in the $PaCO_2$ level ranging from 30–40 mm Hg, and absent in the range of 20–30 mm Hg, suggesting that hyperventilation to levels of $PaCO_2$ less than 30 mm Hg during barbiturate treatment do not effectively increase the degree of vasoconstriction (Kassell *et al.* 1981).

Studies of hypoxaemia in mice show that barbiturate improves survival and decreases mortality (Arnfred and Secher 1962), and *in vitro* studies of slices from hippocampus tissue exposed to periods of hypoxia have shown that the addition of thiopentone to the perfusion medium increases the duration of hypoxia that CA 1 pyramidal cells can survive (Aitken and Schiff 1986). The survival time in hypoxic mice is increased by mild hypothermia and decreased by hyperthermia, and when mild hypothermia is combined with pentobarbitone, survival time generally was further increased (Artru and Michenfelder 1981). This protective effect of barbiturates is bound to their metabolic effect and seems distinct from the anticonvulsant effect (Steen *et al.* 1979). It has also been argued, however, that the protective effect of barbiturate in ischaemia is due to an increase in intracellular pH through the decreased production of lactate (Nordström *et al.* 1978). As judged by metabolic criteria, i.e. measurements of cerebral tissue ATP, phosphocreatine, and lactate in rats subjected to unilateral carotid ligation, hypothermia offers better cerebral protection than does phenobarbitone at the same level of oxygen consumption (Hägerdal *et al.* 1978).

In cats subjected to ICP hypertension by inflation of an epidural balloon, the increase in ICP during inflation was less in animals pretreated with barbiturate than in untreated animals. In the postdeflation period the untreated animals developed a higher ICP increase than did pretreated animals suggesting a protective effect against postcompression brain swelling (Bricolo *et al.* 1981). By following intracellular brain pH in monkeys subjected to MCAO in either halothane or thio-

pentone anaesthesia, it has been shown that the decrease in pH is less pronounced in barbiturate anaesthetized animals. After CBF was restored brain pH returned toward normal after thiopentone anaesthesia, but continued to deteriorate during halothane anaesthesia (Anderson and Sundt 1983).

Several experimental studies of focal and global cerebral ischaemia in cats, dogs, and monkeys indicate that pre-treatment with a barbiturate offers a brain-protective effect by improving survival and diminishing neuropathological changes (Hoff *et al.* 1973, Smith *et al.* 1974, Hankinson *et al.* 1974, Hoff *et al.* 1975, Corkill *et al.* 1976, Safar *et al.* 1976, Michenfelder *et al.* 1976, Corkill *et al.* 1978, Bleyaert *et al.* 1978, Todd *et al.* 1982). In rats subjected to asphyxia and arterial hypotension (Nilsson 1971) and in dogs subjected to hypoxia (Michenfelder and Theye 1973), barbiturate preserves the cerebral energy charge and the lactate production is reduced. Furthermore, in gerbils subjected to unilateral carotid ligation for one hour, neuropathological examination in animals allocated to thiopentone injected one hour after removal of the arterial clamp revealed less extensive neuronal ischaemic cell changes (Levy and Brierley 1979). However, other experimental studies in dogs and monkeys subjected to complete global ischaemia have shown that a barbiturate does not protect the brain (Steen *et al.* 1979, Snyder *et al.* 1979, Gisvold *et al.* 1984, Koch *et al.* 1984), and under these circumstances barbiturate (Steen *et al.* 1978) induces a very moderate if any delay in the depletion of cerebral energy reserves (Nordström and Siesjö 1978). Thus, experimental data do not support the use of barbiturates in the treatment of global ischaemia (Shapiro 1985).

In the discussion concerning the protective effect of barbiturates, the role of sufficient circulation in the reperfusion period must be stressed. In baboons and dogs subjected to temporary MCAO (Selman *et al.* 1981, Yonas *et al.* 1981) or extracranial-intracranial bypass operations (Spetzler *et al.* 1982), sufficient reperfusion after focal cerebral ischaemia is of utmost importance in the protection against cerebral ischaemic damage.

It has been postulated that free radicals are released in ischaemic cerebral tissue, giving rise to irreversible tissue damage (Michenfelder *et al.* 1976). Barbiturates should act as a free radical scavenger thus have a protective effect (Smith *et al.* 1974, Flamm *et al.* 1977). However, this hypothesis has recently been rejected (Smith *et al.* 1980, Siesjö 1984) and at present it is felt that the protective effect of barbiturate is caused by the suppression of cerebral metabolism (Astrup 1982, Siesjö 1984).

Human Investigations

The experimental data concerning the effect of barbiturate on cerebral circulation and metabolism have been confirmed in clinical studies. Thus, barbiturate induces an associated decrease in CBF and CMRO$_2$ and an increase in CVR (Wechsler *et al.* 1951, Pierce *et al.* 1962, Herrschaft *et al.* 1975); and these effects are associated with a decrease in ICP (Søndergard 1961).

In patients with severe head injury with persistent ICP hypertension in spite of hyperventilation, thiopentone induces a rapid decrease in ICP accompanied by an increase in CPP. Sustained ICP reduction could be maintained for several days by combining thiopentone therapy and hypothermia to 30 °C. (Shapiro *et al.* 1974). Other clinical studies support these findings (Miller 1979, Sidi *et al.* 1983). In several uncontrolled clinical studies of patients with severe head injury barbiturate sedation together with hyperventilation have been suggested in the control of ICP hypertension, and this regime has been claimed to improve outcome (Marshall *et al.* 1979 b, Saul and Ducker 1982). Uncontrolled (Rockoff *et al.* 1979) and controlled studies (Eisenberg *et al.* 1988) seem to support these views. On the other hand, uncontrolled studies (Yano *et al.* 1986) and controlled studies (Ward *et al.* 1985) have not confirmed these findings.

Mannitol

In patients with intracranial hypertension, early studies of intravenous infusion of mannitol indicated that this drug effectively reduces ICP (Wise and Chater 1962). After fast intravenous infusion of 0.5–1 g/kg, ICP will be reduced after 2–5 min and the ICP reducing effect lasts for hours, depending on dose and infusion rate (James 1980). The osmotic effect of mannitol is dependent on the osmotic gradient in the blood (Shenkin *et al.* 1962). A difference in osmotic gradient exceeding 10 mOs will give rise to a reduction in ICP (Marshall *et al.* 1978), The decrease in ICP is correlated to the decrease in the water content of brain tissue (Nath and Galgraith 1986).

Animal Experiments

Immediatly after mannitol infusion in dogs (first 2–3 min), ICP will increase. This effect is correlated to an increase in CBV. The increase in CBV persists for 15 min after mannitol infusion, while the ICP returns to control levels within five minutes and continues to decrease (Ravussin *et al.* 1985). Abou-Madi *et al.* (1983) demonstrated that the initial ICP increase can be attenuated when hypocapnia is simultaneously induced. Recent studies from the same group have shown that the initial ICP increase after mannitol is elicited only in animals with a normal ICP but absent in animals with intracranial hypertension (Abou-Madi *et al.* 1987).

In studies of central haemodynamics in rabbits and dogs, mannitol infusion increases blood volume, CVP, wedge pressure and cardiac output, whereas the concentration of haemoglobin, plasma-natrium, and the peripheral resistance decrease. The blood pressure changes, depending on the change in cardiac output and peripheral resistance, and a fall in MABP is often observed (Cote *et al.* 1979, Katz *et al.* 1986, Ravussin *et al.* 1986 a).

Following mannitol infusion in cats, blood viscosity decreases immediatly, the greatest decrease occurring at 10 min. At 75 min a rebound increase in viscosity has been observed. Simultaneously, the diameter of the pial vessels decreases. The largest decrease in diamater was at 10 min. The changes in pial vessels were interpreted as an autoregulatory process (Muizelaar *et al.* 1983). Further studies indicate that if autoregulation is impared, mannitol infusion is followed by a decrease in ICP, whilst ICP remains unchanged in cats with intact cerebral autoregulation (Muizelaar *et al.* 1984). One possible explanation for these findings is that autoregulation is mediated through alterations in the level of adenosine in response to changes in the oxygen availability in cerebral tissue (Winn *et al.* 1980). the decrease in blood viscosity after mannitol administration is thought to improve oxygen transport to the brain. When autoregulation is intact, more oxygen leads to a decrease in the adenosine level resulting in vasoconstriction. The decreased resistance to flow owing to decreased blood viscosity is balanced by an increased resistance due to vasoconstriction. CBF thus, remains unchanged. In spite of the unchanged CBF the increase in resistance gives rise to a decrease in CBV and enhances the osmotic dehydrating effect in ICP. When autoregulation is abolished vasoconstriction does not occur in response to increased oxygen avialability. Owing to the decrease in visocity an increase in CBF occurs. With a lack of vasoconstriction the effect on ICP through dehydration is not enhanced, and the decrease in ICP is much smaller (Muizelaar *et al.* 1984). This hypothesis is based on the assumption that changes in blood viscosity are accompanied by compensatory changes in the degree of constriction in cerebral vessels but, in studies of vessel diameters with the cranial window technique in cats it was concluded that mannitol in clinically relevant doses does not significantly constrict cerebral vessels. Mannitol must therefore exert its effect on ICP through its osmotic effect, rather than by a direct effect on cerebral blood volume (Auer and Haselsberger 1987). Accordingly, studies of mannitol infusion at normal intracranial pressures in baboons (Johnstone and Harper 1973) and dogs (Kassell et al. 1982) have shown an increase in CBF, whereas CBF is unchanged in animals subjected to intracranial hypertension by inflation of an epidural balloon (Johnstone and Harper 1973). In the same study mannitol infusion was followed by an increase in CMRO$_2$.

Mannitol has been shown to have a beneficial effect on experimental cerebral ischaemia (Littler 1978, Watanaba et al. 1979). In studies of experimental cytotoxic oedema, mannitol induces a normalization of the EEG (James et al. 1978). Moreover, studies in rabbits subjected to MCAO have shown that mannitol improves CBF in regions of ischaemia, and the gradual decline in intercellular pH is prevented (Meyer et al. 1987). Using the same ischaemic model in cats, an improved postischaemic recovery of blood flow was observed in mannitol treated animals (Tanaka and Tomonaga 1987), and in rats subjected to forebrain ischaemia, mannitol ameliorated the ischaemic injury (Sutherland et al. 1988). However, other experimental studies of cerebral ischaemia have failed to demonstrate any enhancement of CBF by mannitol (Seki et al. 1981, Pena et al. 1982), and mannitol treatment in monkeys subjected to MCAO failed to improve clinical status or decrease infarct size (Pena et al. 1982).

Human studies of central haemodynamics in patients undergoing craniotomy have shown that mannitol infusion is followed by an increase in blood volume, CVP, wedge pressure and cardiac output and a decrease in the concentration of haemoglobin, plasma-sodium, and the peripheral resistance (Rudehill et al. 1983, Brown et al. 1986). Other studies show an increase in CBV a few minutes after mannitol infusion (Ravussin et al. 1986 a), and following mannitol infusion blood viscosity will decrease for at least two hours. This effect is associated with an enhancement of the cerebral microcirculation (Burke et al. 1981). Accordingly, studies of cerebral circulation have shown an increase in CBF occurring after 10–20 min and lasting for up to 24 hours (Bruce et al. 1973, Mendelow et al. 1985, Jafar et al. 1986) and a variable increase in CMRO$_2$ (Bruce et al.

1973, Jafar et al. 1986). During operation for cerebral tumours or aneurysms, mannitol infusion in patients with normal ICP is followed by a transient, but significant increase in ICP. In contrast, patients with intracranial hypertension showed no increase in ICP, which decreased immediately after mannitol infusion (Ravussin et al. 1986 b). The effect of mannitol on ICP depends on cerebral autoregulation. Mannitol will only decrease ICP effectively in patients with an intact cerebral autoregulation (Muizelaar et al. 1984). Other studies suggest that the effect of mannitol is at least partly dependent upon other haemodynamic mechanisms. Patients with CPP > 70 mm Hg responded relatively poorly to mannitol, whereas ICP decreased in patients with CPP < 70 mm Hg. These findings suggest that at CPP > 70 mm Hg, vasoconstriction already is close to maximum, and therefore mannitol is unable to increase resistance further (Rosner and Coley 1987). In neurointensive patients the volume pressure relationship improves after mannitol often without change in the ICP (Miller and Leech 1975).

Mannitol and Blood-Brain Barrier (BBB)

Osmotic opening of the BBB by infusion of hyperosmolar solutions like mannitol has repeatedly been observed. It has been argued that opening of tight-junctions is the dominant mode of leakage in hyperosmolar opening. The opening of the barrier is independent of energy-producing metabolism. It is thought that osmotic barrier opening is the result of a passive shrinkage of endothelial cells and the surrounding tissue (Greenwood et al. 1988). As a result of the impaired barried function, mannitol diffuses into the cells. Furthermore, brain cells are able to create osmotic active particles. Both factors reduce the transcellular osmotic gradient (Feig and McCurdy 1977, Jennett and Teasdale 1981). After dicontinuing the mannitol infusion, the osmotic gradient is reversed because mannitol is excreted in the urine. Consequently, the concentration of mannitol is found to be higher in the intracellular compartment as compared with the concentration in the extracellular space. The change in osmotic gradients is followed by influx of water into the brain cells and an increase in ICP (McQueen and Jeanes 1964). The occurrence of this *rebound phenomenon* indicates that mannitol primarily should be used in a single dose in the treatment of ICP-hypertension.

In rats, the duration of BBB damage after mannitol infusion lasts 24 to 48 hours (Suzuki et al. 1988). In experimental studies neuronal damage has been associated with **BBB** dysfunction after mannitol infusion (Rapoport *et al.* 1972, Tomiwa *et al.* 1982). Two distinct types of cerebral tissue damage have been identified; microinfarction (cerebral neocortex, hippocampus, thalamus, midbrain and cerebellar hemisphere) and "ischaemic neuronal death" of delayed onset (CA-1 region of hippocampus, the cerebellum, and the thalamus). The high incidence of neuronal damage has been attributed to embolization of cerebral vessels with precipitates of mannitol and a compromized blood flow in the carotid artery during mannitol infusion (Suzuki *et al.* 1988).

4. Experimental Studies of Head Injury

The dynamic pathophysiological changes which follow head injury are caused by successive events. First, the trauma directly influences the cerebral circulation and metabolism. Later, the effect of mass-lesions caused by haematoma, ischaemia, brain oedema, intracranial hypertension, and systemic hypotension and hypertension may influence ICP, CBF, and $CMRO_2$. Accordingly, studies in experimental models, including one or more of these pathogenic factors, are important in the understanding of the dynamic changes following head injury.

Experimental Studies

Four models of acute head injury have gained considerable interest. The non-impact acceleration device model of Gennarelli (1983), the fluid percussion model in cats (Sullivan *et al*. 1976), the impact acceleration model in rats (Nilsson *et al*. 1977 a), and the Remington stunner model in cats (Tornheim *et al*. 1981). In these models a good correlation between trauma intensity and the pathophysiological and biochemical changes in the cerebral tissue has been found (Nilsson *et al*. 1977, Tornheim *et al*. 1983, Wagner *et al*. 1985, Gennarelli *et al*. 1986). The depletion of ATP and phosphocreatine and the increase in lactate are thus related to the intensity of the trauma. In some experiments the increase in lactate caused by the trauma is unevenly distributed within the brain, the hippocampus being the most vulnerable part of the brain (Yang *et al*. 1985). The increase in cerebral lactate concentration is of interest, because brain tissue lactic-acidosis may be an important factor in the development of cell damage (Siesjö and Wieloch 1985). Furthermore, prevention of the spread of local acidosis to adjacent normal tissue may have a beneficial effect on the recovery of brain tissue (Marshall *et al*. 1975, Becker *et al*. 1979). Accordingly, experimental studies have shown that treatment of brain acidosis with the weak base tromethapine reduces the incidence of intracranial hypertension and decreases the mortality in cats subjected to controlled ventilation over a 3 day period following fluid percussion injury (Rosner and Becker 1984).

Gennarelli (1983) used a single *non-impact controlled acceleration device* which delivered movement of the head in a single plane. In experimental studies in monkeys the direction of the acceleration was found to be of importance. The sagital motions produced greater ICP elevation than did coronal plane motions (Gennarelli *et al*. 1983). In recent studies the same group demonstrated that for an equal degree of acceleration, coronal plane motion produced less ICP change than did sagital motion intersperced with horizontal motion and that ICP hypertension developed with increasing severity of injury from concussion to diffuse axonal injury (Gennarelli *et al*. 1986).

An *impact acceleration trauma* of 7–9 m/sec in rats is followed by an increased metabolic rate and increased neuronal activity (Nilsson and Nordström 1977 a and b, Nilsson *et al*. 1977). The depletion in brain energy metabolism occurred at 1 min and was most pronounced at 4 min. Restitution was found after 15 min. At an acceleration trauma of 9 m/sec the metabolic changes were most pronounced in the brain-stem (Nilsson and Pontén 1977). Immediately after the trauma CBF was increased. However, within few min a 30–40% decrease was observed and normalization of CBF occurred over 20–30 min. $CMRO_2$ increased during the first minutes. Later, a decrease in $CMRO_2$ was observed. This decrease correlated with the intensity of the trauma (Nilsson and Nordström 1977 a).

In cats subjected to *fluid percussion* and studied with the cranial window technique, CBV was increased after the blow (Duckrow *et al*. 1981). However, in the same model, CBF was relatively resistant to the blow (DeWitt *et al*. 1981, Lewelt *et al*. 1982). Using a microsphere technique in a similar model, DeWitt *et al*. (1986) observed that within one minute of injury an increase in CPP, dilatation of the arterioles, and an increase in CBF occurred. In the same study no evidence of reduced CBF was found in any region studied. These changes returned to preinjury levels by 30 min.

In contrast, recent studies of fluid percussion in piglets indicated a decrease in CBF and CPP and an increase in ICP within 5 min after the trauma (Pfenninger *et al.* 1989 b).

Recently MR spectroscopy has been used in the study of rats subjected to fluid percussion trauma of 1.5–2.5 atmospheres. Following the injury the ratio of phosphocreatine to inorganic phosphate (PCr/P) showed a biphasic decline. The first decline reached nadir at 40 minutes with recovery at 100 min. At two hours this event was followed by a second decline that persisted for the remaining six hours. The first but not the second decrease in PCr/P ratio was associated with tissue acidosis. These changes may indicate an altered mitochondrial energy production and not be caused by CBF reduction (Vink *et al.* 1987). Similar studies by Hashimoto *et al.* (1986) have shown that a change in phosphorus metabolism occurs over a period of time (15–20 min). In the same study it was demonstrated that ATP peaks were preserved until phosphocreatine was almost entirely depleted. Studies in cats subjected to fluid precussion trauma indicate that the excess of lactate and decrease in tissue pH are not accompanied by depletion of ATP or phosphocreatine and that the level of CBF, $CMRO_2$, and CMR-glucose are relatively unchanged. Under these circumstances the authors conclude that flow deprivation does not cause cerebral ischaemia. Theoretically, these changes might indicate a mild derangement of brain energy metabolism reflecting altered mitochondrial function (Yang *et al.* 1985, Unterberg *et al.* 1988). However, when cerebral percusion trauma was combined with hypoventilation in order to mimic true cerebral contusion, a marked alteration in CBF, CMR-glucose, PCr/P ratio, and tissue pH were observed indicating cerebral ischaemia (Andersen *et al.* 1988). Studies with electron microscopy in rats subjected to brain concussion resulting in unconsciousness for 3–10 min have shown swelling of the mitochondria, beginning after 30 min, with maximum after one hour, and disappearence after 24 hours (Bakay *et al.* 1977).

The CA is easily abolished by fluid percussion trauma and the degree of impairment is correlated to the intensity of the trauma (Lewelt *et al.* 1980, Seelig *et al.* 1983). In experiments in cats the trauma was found to attenuate the capacity of CBF to increase during hypoxaemia (Lewelt *et al.* 1982). In rats subjected to impact of 4.9 atmospheres plus hypoxaemia to PO_2 40 mm Hg for 30 min, increased morbidity was associated with more widespred and severe cerebral oedema and a decrease in CBF in the ipselateral hem-

isphere (Ishige *et al.* 1987). Other studies in rats indicate that the brain is extremely vulnerable to hypovolaemic hypotension, as reflected by the loss of high-energy phosphates in the brain (Ishige *et al.* 1988).

In cats subjected to fluid percussion, CO_2 reactivity is inversely related to the severely of the trauma (Saunders *et al.* 1979).

The effect of a *mass-expanding lesion* on CBF has been studied experimentally with subdural or epidural balloons (Johnston *et al.* 1973, Goodman and Becker 1973, Sullivan *et al.* 1977). In these models the microscopic and macroscopic changes are related to the expansion of the mass lesion (Goodman and Becker 1973). CBF is well maintained as long as CPP is preserved above the lower level of CA (Johnston *et al.* 1973). No pressure gradients were present between supratentorial compartments, but pressure differences occurred between infratentorial and supratentorial regions (Johnstone and Rowan 1974). With an increase in mass effect a progressive increase in the pressure-volume index has been observed, indicating compromized cerebral compliance (Sullivan *et al.* 1977).

The influence of increased ICP on CBF also been studied in mock *CSF infusion models*. In dogs a decrease in CBF has been observed when CPP falls below 60 mm Hg and a further decrease in flow and $CMRO_2$ take place at CPP values of 40 mm Hg. The changes in CBF follow the principles of regulation of the CA (Hamar *et al.* 1973). In baboons subjected to mock CSF induced ICP hypertension, CBF remained constant until ICP levels of 50 mm Hg; in the interval of ICP 50–100 mm Hg a marked increase in CBF was associated with systemic hypertension. A further increase in ICP resulted in a progressive decrease in CBF. Prior section of the spinal cord prevented the increase in CBF and the systemic hypertension (Johnston *et al.* 1972). In dogs subjected to CPP reduction to 30–35 mm Hg by mock CSF infusion, depletion of ATP and phosphocreatine and an increase in cerebral lactate were observed. Resting energy states were not approached until 30–90 min after restitution of normal CPP. In the same study CSF pressure release was followed by hyperaemia of approximately the same duration (Zwetnow 1970).

As cerebral ischaemia is a frequent finding in patients dying after head injury, some experimental findings in complete and incomplete ischaemia will be discussed. In rabbits Marshall and coworkers (1975 b) induced *incomplete* (CPP 20 mm Hg), and *complete ischaemia* of 15 min duration. They found normal brain morphology in oligaemic animals but ischaemic damages in the striatum and hippocampus in complete ischaemia. In this model high energy metabolism was exhausted within 5 min of the insult. Recovery of ATP and phosphocreatine occurred within 15 min and before EEG activity was observed. Severe lactic acidosis persisted for at least 15 min after the insult (Marshall *et al.* 1975 c). During the oligaemic period CBF was about 20% of contral values, but hyperaemia was seen from 1 to 5 min after reestablishment of cerebral circulation. Oligaemic animals had a better outcome and hypocapnia improved neurological outcome (Marshall *et al.* 1975 a).

In the immediate postischaemic period $CMRO_2$ is reduced (Snyder *et al.* 1975, Hossmann *et al.* 1976, Nordström and Rehncrona 1977). Later, $CMRO_2$ increases toward control values and under certain circumstances $CMRO_2$ might overshoot the control value (Hossmann *et al.* 1976, Kofke *et al.* 1979, Levy and Duffy 1977). Studies in cats with MR spectroscopy during and after a period of hypoxia or ischaemia indicate that an increase in lactate is accompanied by a decrease in intracellular pH. Furthermore, a delayed recovery of phosphocreatine has been found to be due to metabolic acidosis and not to delayed recovery of ATP and ADP (Gyulai *et al.* 1987). In experiments with repeated ischaemia insults in gerbils, a pronounced cummulative effect on the development of cerebral oedema was observed (Tomida *et al.* 1987). Siesjö (1984) has reviewed the changes in cerebral circulation and metabolism during and after incomplete and complete ischaemia, and recently the same author suggestss that elective neoronal vulnerability observed after brief periods of ischaemia is elicited by postsynaptic damage to neurons innervated by exitatory amino acids (aspartate and glutamate) which give rise to enhanced calcium influx and osmolytic damage (Siesjö 1988).

Cerebral oedema is frequent finding in experimental as well as human head trauma. The development of oedema has been studied experimentally after cold injury. In cats this lesion elicites a focal CBF decrease and rCBF was found to be inversely proportional to the water content in the brain tissue (Frei *et al.* 1973). The decrease in rCBF is attenuated by hyperbaric oxygen (Miller *et al.* 1970). Local glucose utilization is decreased in thermally traumatized rat brains (Pappius 1981). Recent studies have shown that pentoparbitone reduces local glucose utilization proportionally with control values obtained before pentobarbitone administration (Elphinstone *et al.* 1988). Tissue pressure differences of 13 mm Hg between oedematous and normal brain tissue have been observed (Reulen and Kreysch 1973). In cats subjected to hemispheric cryogenic injury, inter-hemispheric pressure gradients reached about 4 mm Hg within one hour; however, the transtentorial pressure gradient remained above 10 mm Hg for several hours (Furuse *et al.* 1981). These differences in pressure promote buld flow and spread of oedema (Reulen *et al.* 1977). Cerebral oedema is enhanced by a sudden increase in MABP which also provoke an increase in ICP (Schutta *et al.* 1968).

In experimental studies of brain trauma evidence has been advanced that release of prostaglandins play a part in the generation of cerebral oedema (Shohami *et al.* 1987). In cats subjected to fluid percussion trauma phospholipase C, a precurser enzyme responsible for the generation of diacyl to prostaglandins, is elevated (Ellis *et al.* 1981, Wei *et al.* 1982). However, the kallikrein-kinin system eliciting brydykinin, also plays a role as mediator of vasogenic oedema (Unterberg and Baethmann 1984, Maier-Hauff *et al.* 1984). Bradykinin provokes massive vasodilatation of cerebral vessels and opening of the BBB (Wahl *et al.* 1983, Unterberg *et al.* 1984). These effects can be prevented by aprotinine, a trypsine inhibitor (Unterberg *et al.* 1986). Thus, experimental studies suggest that prostaglandins as well as bradykinin are responsible for the formation of cerebral oedema. Moreover, superoxide (Kontos and Wei 1986) and vasopressine (Reeder *et al.* 1986) may also be responsible for the development of cerebral oedema after head trauma.

5. Human Studies of Head Injury

Neuropathological studies of brains from consecutive patients dying as a result of blunt head injury have shown obvious signs of ischaemic damages in 91%. These changes are found in the cerebral cortex in 46%, hippocampus 81%, cerebellum 36% and to the basal ganglia in 79% (Graham *et al.* 1978). Moreover, direct impact after a blunt lesion and secondary events like hypotension might result in localized necrosis at boundary zones between the major cerebral arteries (water shed lesions) (Adams *et al.* 1966, Adams and Graham 1976); furthermore, neuropathological findings of brain herniation suggest well-defined ischaemic lesions of parahippocampal and cingulate gyri, and infarction in the medial occipital cortex (Adams and Graham 1976). The ischaemic brain damage is probably due to intracranial hypertension caused by mass lesions and brain swelling (Langfitt and Gennarelli 1982).

In human, impacts associated with head injury may have a translational and rotational component. Passing through the cerebral tissue the pressure wave disrupts neuronal axons, gives rise to a diffuse brain lesion, and provokes cerebral contusion immediatly below the impact, and at the uppersite side of the trauma (contrecoup). In accordance with experimental studies by Ommaya and Gennerelli (1974), acceleration trauma with rotational component produces centripetal progression, which is always maximal in the peripheral region. Compared with subcortical regions the cerebral damage therefore is more severe in the cerebral cortex and the authors argue that a primary brain-stem lesion does not excist. Neuropathological studies in humans support this view (Mitchell and Adams 1973). Consistent with the centripetal model of progressive brain injury Levin *et al.* (1988) found that the depth of the brain lesion, evaluated by MR imaging, was positively related to the degree and the duration of impaired consciousness.

ICP Recordings in Head Injury

Several investigations of ICP in patients with severe head injury indicate that intracranial hypertension (IH) often occurs. Pressure levels above 30–40 mm Hg for several hours is associated with a poor outcome (Troupp 1967, Vapalahti *et al.* 1969, Cold *et al.* 1975, Changaris *et al.* 1987). In a study of 160 patients with severe head injury Miller *et al.* (1977) found IH (ICP > 10 mm Hg) in 82% of the patients and in 97% of the patients with a mass lesion. Although ICP > 40 mm Hg was only observed in 10% of the patients, IH was found to be the primary reason for death in 50% of the patients. Recently these findings were confirmed by the same investigators (Miller *et al.* 1981). In a study of severe head injury where carotid blood flow was determined by electromagnetic flowmeters, CBF was found to change passively with changes in blood pressure indicating total loss of CA. In the same study ICP levels above 40 mm Hg were associated with flow deprivation (Gobiet *et al.* 1975). In the final stages of severe HI incarceration is accompanied by a characteristic pattern of sudden ICP rise (Troupp and Vapalahti 1971) and a decrease in CPP to very low levels (Balslev-Jørgensen *et al.* 1972).

Based on clinical studies several investigators have argued that ICP should be recorded as it provides a guide to therapy (Johnston *et al.* 1970, Becker *et al.* 1977, Papo 1977, Marshall *et al.* 1979 a, Marshall *et al.* 1979 b, Kobayashi *et al.* 1983, Moss *et al.* 1983). Moreover, a high level of ICP is correlated to a poor outcome (Kobayashi *et al.* 1983, Klauber *et al.* 1984). Furthermore, it has been demonstrated that ICP monitoring might be a guide to aggressive therapy including prolonged hyperventilation, osmotic therapy and barbiturate coma treatment (Becker *et al.* 1977, Marshall *et al.* 1979 b, Saul and Ducker 1982). In this context studies by Rosner and Coley (1986) are of interest indicating that a low CPP might precipitate ICP-hypertension. Rosner argues that a low CPP is capable of stimulating cerebral vasodilatation and increase CBV and ICP (vasodilatory cascade). The same author has demonstrated that a MABP decrement often precedes ICP pressure waves. These findings suggest that CA might be present even in the acute phase of HI, and the author argues that the optimal CPP is close to 85–90 mm Hg in normotensive patients, at which level the lowest level of ICP is observed (Rosner 1987).

The significance of cerebral oedema in HI has been evaluated by Galbraith et al. (1984). They found a striking correlation between high water content in white matter around an intracranial haematoma (determined by gravimetric technique in biopsies taken peroperatively) and the level of ICP. In patients with a diffuse cerebral lesion this correlation was not observed. Mendelow et al. (1983) calculated CBV by CT scanning taken before and after intravenous injection of iodine contrast. They found CBV ranging from 2–8%. An association between high CBV and elevated ICP was not found. In another study in patients with HI, ventricular volume was measured by CT scanning before and after evaculation of an intracranial haematoma. A correlation between a decrease in ventricular volume and an increase in ICP was observed (Murphy et al. 1983). Other studies including preoperative volumetric analysis of the ventricles in patients with extradural haematoma have shown that an increase in the total volume of the ventricles above normal, a midline shift exceeding 10 mm and obliteration of the third ventricle, and basal cisterns might contribute to the increase in ICP (Findley et al. 1983). In patients with diffuse injury no relationship between the size of the ventricles and ICP has been found; however, an association between obliteration of the third ventricle and the basal cisterns and an increase in ICP was observed (Teasdale et al. 1984). In this context a study by Marmarou et al. (1987) should be mentioned which concluded that CSF absorption parameters (CSF-formation and outflow resistance) accounted for approximately one-third of the ICP rise after severe HI. The authors suggest that other mechanisms presumably vascular in origin might be a predominant factor in elevation of ICP. Recently, ICP monitoring in patients with severe HI has been discussed. ICP recording is found to be indispensable in patients with an abnormal CT scan (mass lesion), patients with posturing, patients over 40 years of age, and patients with a period of arterial hypotension (Narayan et al. 1982, Becker et al. 1983, Teasdale et al. 1986).

Several studies indicate that the outcome in paediatric patients is better than adults (Bruce et al. 1978). However, no difference in outcome in paediatric and adult patients was found in patients with a mass lesion and patients with increased ICP, regardless of whether ICP was reducible (Alberico et al. 1987). In children IH is especially found in patients with Glasgow coma score < 5. It is concluded that aggressive therapy with ICP monitoring in order to keep ICP < 20 mm Hg seems justified (Bruce et al. 1978).

The level of ICP indicating antihypertensive therapy is debated. ICP of about 20–25 mm Hg is tolerated for short periods of 5–10 min; however, pressures above this level for longer periods (20–30 min) suggest re-evaluation as regards respiration, circulation, degree and choice of sedation, and neurological evaluation with special reference to the occurrence of an expanding lesion. If HI persists and operative intervention is not indicated, the available therapeutic measures include controlled hyperventilation, sedation with barbiturates or other hypnotic agents, the use of analgetics (provided ventilation is controlled), and osmotic therapy. Moreover, ventricular drainage in selective cases is effective in controlling IH (Johnston et al. 1970); however, continuous drainage for longer periods might increase the compression of the cerebral ventricles and thus negate the effect of drainage.

The value of the pressure-volume index (PVI) has been evaluated by Tans and Poortvliet (1983), who concluded that a PVI index < 10 ml indicates antihypertensive treatment. Maset et al. (1987) found a clear relationship between low PVI values measured soon after the injury and subsequent development of ICP-hypertension. However, in other studies PVI varied independently of the clinical course, outcome, and ICP, making any decisions regarding therapeutic interventions difficult (Kostaljanetz 1986, Marmarou et al. 1986). As stated by Gray and Rosner (1987), in experimental studies the PVI is a complex function of CPP varying directly with CPP within the autoregulatory range and indirectly with CPP below the autoregulatory range.

Biochemical Studies of CSF

In clinical studies of acute HI a high concentration of lumbar CSF lactate (Kurze et al. 1966, Zupping 1970, Crockard and Taylor 1972, King et al. 1974, Sood et al. 1980), suboccipital CSF lactate (Metzel and Zimmermann 1971, Seitz and Ocker 1977) and ventricular fluid lactate (Fieschi et al. 1974, Overgaard and Tweed 1974, Cold et al. 1975 b, Enevoldsen et al. 1976, Enevoldsen and Jensen 1978 a) have been observed during the first days after trauma. Generally, an initially or persistantly high lactate level indicates a poor outcome. However, in patients with severe head injury the interpretation of increased CSF lactate is difficult owing to contamination by blood (Granholm 1969), hypocapnia (Granholm and Siesjö 1969, Cold et al. 1975) and CNS infection all of which will increase CSF lactate. However, in patients with severe HI subjected to controlled ventilation with $PaCO_2$ ranging from 34–40 mm Hg a high concentration of lactate in ventricular fluid is of prognostic value (Cold *et al.* 1975 b. Cold 1981). In a recent study of patients with head injury subjected to moderate hypocapnia, the level of ventricular fluid lactate during the first four posttraumatic days was significantly and persistantly higher in patients with a poor outcome; in contrast, the concentration in patients with a good outcome declined rapidly within 48 hours. In the same study lactic acidosis was found to be associated with a high ICP level (DeSalles *et al.* 1986). In another study the increase in ventricular fluid lactate correlated to a decrease in extracellular pH (DeSalles *et al.* 1987 b). In cats subjected to fluid percussion trauma the concentration of lactate in brain tissue, CSF, and blood increased in proportion to the severty of the injury. The initial elevation of CSF lactate may reflect the systemic response to trauma; the secondary rise of CSF lactate levels may be due to slow seepage of lactate produced by brain tissue into the CSF space (Iano *et al.* 1988). Short term hyperglycaemia related to stress after the injury coupled with an impaired cerebral oxidative metabolism is supposed to be the reason for the initially high lactate level (Clifton *et al.* 1981, Clifton *et al.* 1983). However, the increase in ventricular fluid lactate outlasts the acute response to injury, and the increase in lactate concentration may reflect a dysfunction in cerebral cellular metabolism caused by the injury itself and aggravated by secondary insults (DeSalles *et al.* 1987 a). In a preliminary study Seitz and Ocker (1977) demonstrated that treatment of CSF acidosis by intrathecal injection of bicarbonate had an improving effect on CBF as well as brain metabolism in patients with severe HI. Clinical studies in patients with severe HI indicate that intravenous treatment with THAM (trishydroxymethylaminomethan) effectively controls ICP-hypertension (Pfenninger *et al.* 1989 a), and when used prophylactically, a preliminary controlled study revealed lower levels of ICP, lower levels of CSF-lactate and probably improved recovery (Rosner *et al.* 1989).

Creatine-kinase-BB isoenzyme activity has been shown to rise very rapidly in CSF after parenchymal brain damage (Maas 1977, Rabow and Hedman 1979, Bakay and Ward 1983). The increase may be due to destruction of cell membranes, with release of the enzyme into the extracellular space and seepage into the CSF. Clinical studies have shown that creatine kinase isoenzyme activity in CSF is of prognostic significance (Kaste *et al.* 1981, Nordby and Urdal 1982, Rabow and Hedman 1985); however, recent studies by Rabow *et al.* (1986) indicate that no significant correlation between the activity of the enzyme and outcome among surviving patients could be obtained. The same investigators found that the creatine kinase activity reached nadir a few hours after the trauma. The activity of creatine kinase had a monoexponential drop with a half-time of approximately 10 hours. Other clinical studies show a negative correlation between the activity of *cyclic AMP* and the coma grade following HI (Fleischer *et al.* 1977).

6. The Author's Own Studies

CBF and CMRO$_2$ in the Acute Phase of Head Injury

In the following nine studies done by the author of CBF and metabolism in the acute phase of HI are reviewed. Repeated studies of CBF in the acute phase after head injury (study V), cerebral metabolic rate of oxygen (CMRO$_2$) in the acute phase of head injury (study III), the relationship between CMRO$_2$ and CBF in the acute phase of head injury (study VII), studies of cerebral autoregulation (study IV), studies of the effect of two levels of induced hypocapnia on cerebral autoregulation (study VI), studies of the CO$_2$ reactivity during the acute phase (studies I, II, VIII, and IX), and studies of the barbiturate reactivity and CO$_2$ reactivity during the acute phase (study IX).

Presentation of the Study and the Investigational Methods Used

The present studies of rCBF and metabolism include 57 patients with severe HI. The investigations were performed in three periods. Fourty patients were studied in the period 1970–73 with a 16 channel Cerebrograph, and rCBF was calculated as initial slope index or by stocastic analysis after intraarterial injection of Xenon-133. In the period 1976–77 8 patients were studied by the intraarterial approach. These investigations included studies of CA during two levels of PaCO$_2$. Finally, nine patients were studied in the period 1986–87 with a 32 channel Cerebrograph using the inhalation method and rCBF was calculated as ISI and as CBF-10.

At admission to the neurosurgical departments all 57 patients were comatose with an estimated Glasgow coma score < 7, median age 23 years (range 7–70). The first studies in the period 1970–73 were performed before CT-scanning was avialable. In these patients the severity and the site of the cerebral lesions were determined by arteriographic studies, findings during operation (explorative burr-holes or craniotomy) and by repeated neurological examination including evalua-

tion of the state of consciousness, and the brain stem reflexes (pupillary, ciliary, corneal, oculocephalic, and vestibular reflexes to cold water). The remaining 17 patients were studied in 1976–77 and 1986–87. In these patients repeated CT scannings were done which supplemented the operative findings and the neurological examinations. On the basis of these investigations the 57 patients were divided into patients with cerebral contusion or subdural haematoma and normal brain-stem reflexes (15 patients); patients without cerebral contusion or subdural haematoma, but with abnormal brain-stem reflexes (6 patients); patients with cerebral contusion or subdural haematoma and abnormal brain-stem reflexes (35 patients); and patients with diffuse cerebral swelling on CT scanning, but with normal brain-stem reflexes (1 patient).

In the first study, 40 patients (period 1970–73), a follow-up study was performed from 6 months to 2 years after the trauma. The patients were divided into two groups. Group I included 22 patients with good recovery or slight mental impairments, insufficient to prevent resumption of previous work or school. Group II included patients with severe mental impairment or dementia (7 patients) and vegetative survivals or patients who died without regaining consciousness (11 patients). In the remaining 17 patients the outcome was evaluated in accordance with Teasdale and Jennett (1976) as good recovery (3 patients), moderate disability (5 patients), severe disability (6 patients), and death (3 patients). The mortality in this material was thus 25%.

After neuroradiological examination and operation all 57 patients were admitted to the intensive care unit of the neurosurgical department. The patients received continuous moderate hyperventilation and appropiate sedation. ICP was continuously recorded in 55 patients: in 42 patients as intraventricular pressure (Lundberg 1960), in 5 patients with an epidural pressure by means of a non-coplanar fiber-optic sensor (Ladd Pressure Monitor). In 8 patients subdural pressure was recorded. During the CBF measurements mean arterial blood pressure (MABP) was continuously monitored. In 35 patients ventricular fluid was repeatedly analysed for lactate, pyruvate (Marbach and Weil 1967), pH, and bicarbonate.

During the first study, which included 40 patients, regional studies of the CO_2 reactivity were performed in 27 patients. CA was tested in 17 patients during angiotensin infusion and $CMRO_2$ was tested in 22 patients. In the period 1976–77 CA was tested in eight patients during two levels of $PaCO_2$. In the last study performed in the period 1986–87, regional studies of CO_2 reactivity were performed in eight patients and global and regional studies of barbiturate reactivity (defined as change in $CMRO_2$ or regionally (change in rCBF) after a bolus injection of thiopental 5 mg/kg) were done in 7 patients.

In the following details of the nine studies will be presented

Repeated Studies of CBF (Study V)

Material: 21 patients (16 males and 5 females), median age 23 years (range 9–67 years), admitted to Neurosurgical Department G, Århus Kommunehospital for severe head injury. Patients with manifest heart, lung and abdominal trauma and patients who after initial evaluation or neurosurgical exploration were in a state of brain death were excluded from the study. The nature of the cerebral injury was evaluated on the basis of the clinical examination including studies of brainstem reflexes, the results of the initial neurosurgical exploration and the neuroradiological examination. On this basis diagnose in the 21 patients were as follows: cerebral contusion or dilaceration without lesion of the brain-stem (2 patients), cerebral cortical contusion or dilaceration with brain-stem lesion (16 patients), and patients without cortical lesions but with brain-stem lesions (3 patients). All patients were subjected to controllend ventilation, the aim being $PaCO_2$ levels of about 4 kPa. Moderate sedation was provided by meperidine 20 mg, diazepam 5 mg and chlorpromazine 5–10 mg i.v.

A follow-up study was performed from 6 months to 2 years after the trauma. The studies involved a complex of psychological tests in order to classify the neurological and mental sequelae. The patients were classified as follows. Group I: Complete recovery or slight mental impairment without dementia (the patients were able to resume their previous work or school (nine patients). Group II: Dementia present, vegetative survival, and patients who died without regaining consciousness (twelve patients).

Regional cerebral blood flow (rCBF) was calculated by initial slope index with the intraarterial 133-Xenon technique. A 16-channel Cerebrograph (model RCBF 1616) was used. Via a magnetic tape recorder the clearance curves were simultaneously displayed semilogarithmically on paper. The counting rates were punched on paper tape using a time constant of 5 s. If the curves were regular and without defect due to drop out of counts during the first two min, a program using regression analyses was used to calculate the initial slope index. When a defect was present, the semilogarithmically curves were used for the calculation. CBF was measured over the most injured hemisphere. In 18 patients, CBF was measured twice. The first CBF study was performed without changing the setting of the ventilator. The second after hyperventilation. In these patients the first CBF was corrected for changes from $PaCO_2$ 4 kPa. In the other studies the average value of CO_2 reactivity (delta ln CBF/delta $PaCO_2$ for group I 0.037 and for group II 0.026) was used for correction. These corrected values were in accordance with data obtained in Study II.

In all patients intraventricular pressure was continuously recorded throughout the study. MABP was recorded intraarterially.

Results: In seven patients, low CBF values (< 30 ml/100 g/min) were found throughout the study. Three of the patients (ages 20, 23 and 45) had mainly brain-stem lesions or diffuse brain lesion; one of these patients recovered completely, and two were classified as group II. The remaining four patients had both cortical and brain-stem lesions. The ages of the four patients were above 45 years. The clinical outcome was good in two (group I) and poor in two patients (group II) (Fig. 4 lower part).

In 4 patients, in one of whom CBF was measured repeatedly on both sides, CBF increased above 30 ml/100 g/min, and the maximal CBF value was observed within 24 hours after the trauma. All patients had cortical lesions of the hemispheres (contusion or dilaceration). In one patient, aged 9 years, a prolonged hyperperfusion state was found. This patient had no brain-stem lesion and recovered completely. In the other three patients (ages 25, 26 and 67 years) brain-stem lesion was present throughout the studies. A normal or low-flow state persisted in two of these patients, who both died. In one patient, who recovered (group I), bilateral studies were performed three times during the first 48 hours. In both hemispheres maximal flow was recorded within the first 24 hours (Fig. 4, upper part).

CBF$_{init.}$

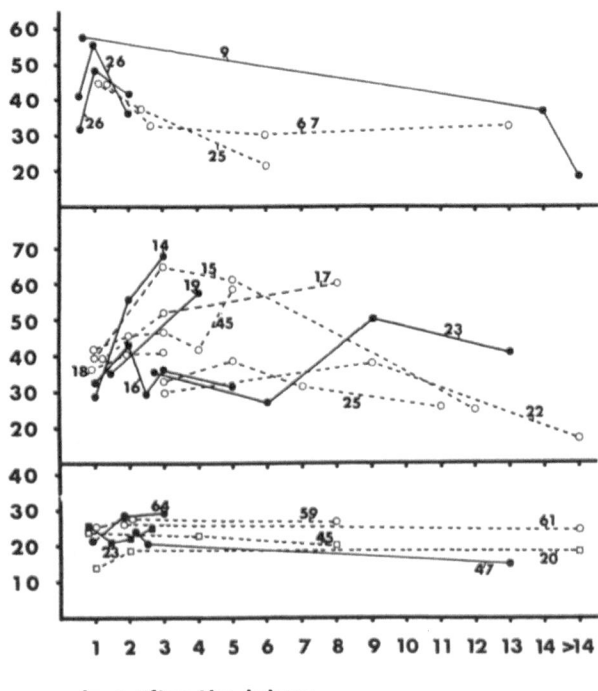

days after the injury

Fig. 4. Repeated studies of the average hemispheric CBF (initial slope), measured over the most severely injured hemisphere in 21 unsconcious patients with head injury. CBF was corrected for changes from PaCO$_2$ 30 mm Hg by simultaneous studies of CO$_2$ reactivity. The ages of the patients are indicated on the figures. Unbroken lines and closed circles indicate good clinical recovery (complete recovery or slight mental impairment). Broken lines and open circles indicate a poor clinical outcome (dementia, vegetative survival and death). Circles indicate patients with contusion or dilaceration. Squares indicate patients with brain-stem lesions without contusion or dilaceration of the hemispheres. *Upper figure* indicates an early hyperperfusion state in four repeated studies, one of which bilateral. *Central figure* indicates a late hyperperfusion state in 10 patients. *The lower figure* indicates a low perfusion state in three patients without hemispheric lesions but with lesion of the brainstem, and in four patients with hemispheric lesions (all with ages above 45 years) (Acta Anaesthesiol Scand 1980: 24: 245–251, with permission)

In 10 patients, the maximal CBF was observed after the first day following the trauma. In all patients CBF increased above 30 ml/100 g/min. The period with hyperperfusion lasted from 1 to 8 days. These patients suffered from lesions of the cerebral hemispheres. Nine of the patients had lesions of the brain-stem. The ages of the patients ranged from 14 to 45 years, and in eight of the patients the ages were below 25 years. Five of the patients recovered (group I); five victims did poorly (group II) (Fig. 4, central figure).

Other results concerning dynamic changes in CBF

In patients with space-occupying lesions (contusion, subdural haematoma, oedema) hemispheric CBF generally was highest ipselateral to the lesion (study IX).

In patients who regained consciousness CBF generally increased (study V).

In patients with intracranial mass-lesions decompression was followed by an increase in CBF (study V).

In patients with reduced hemispheric CBF (mean CBF < 30 ml) the mean value of AVDO$_2$ was higher and the mean value of ventricular fluid pH was lower (study VI).

Studies of CMRO$_2$ (Study III)

Material: 14 patients (9 males and 5 females); median age 23 (range 7–67) were studied. Throughout the study all patients were unconscious Glasgow coma score < 7.

In accordance with study 1 the patients were divided into two groups. Group 1 included 7 patients; group II 7 patients.

In accordance with study V CBF was measured over the most severely injured hemisphere. CBF was calculated as stocastic flow (height-over-area) using the 15 min clearance curves. The mean CBF was calculated as the average of the 16 rCBF values. The jugular bulb was punctured from the lateral approach, and AVDO$_2$ was calculated as the difference between oxygen content in arterial and venous blood. CMRO$_2$ was cal-

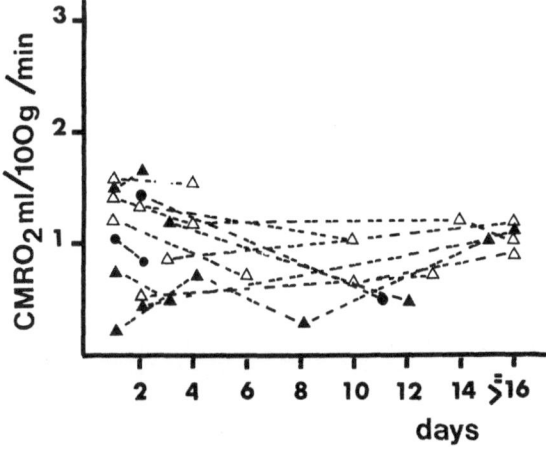

Fig. 5. Cerebral metabolic rate of oxygen (CMRO$_2$) related to time after the head injury in 14 patients in whom repeated studies over the most injured hemisphere were performed. Open triangles indicate patients with a good clinical recovery (good recovery or slight mental impairment). Closed triangles indicate patients with severe impairments or dementia. Closed circles indicate patients with vegetative survival or death (Acta Anaesthesiol Scand 1978: 22: 249–256, with permission)

culated as the product of mean CBF and $AVDO_2$. During the CBF study the patients were sedated as in study V.

Results: $CMRO_2$ ranged between 0.24 and 2.21 ml $O_2/100$ g/min. $CMRO_2$ was not related to clinical outcome or the presence or absence of brain-stem reflexes. Repeated studies of $CMRO_2$ spaced days apart indicated that $CMRO_2$ is constantly reduced in unconscious patients. A critical low $CMRO_2$ was not defined, and $CMRO_2$ at 0.4 ml O_2 was compatible with restitution of intellectual function.

Other results concerning dynamic changes in $CMRO_2$:

$CMRO_2$ was negatively correlated to ventricular fluid lactate (r $= -0.434$, P < 0.05) (study III).

Studies of Relationship Between $CMRO_2$ and CBF (Study VII)

Material: 16 unconscious patients (Glasgow coma scale < 7), median age 23 (range 9–67 years). The patients were on continuous respiratory treatment and sedated in accordance with the principles of study V.

CBF and $CMRO_2$ were calculated in accordance with study II. The interval between the repeated studies (2–4 studies per patient) was 1 day in five studies, 2–4 days in six studies, 5–7 days in four studies and 8–16 days in six studies. Correction of CBF for changes in $PaCO_2$ was performed with a 3% change in CBF per

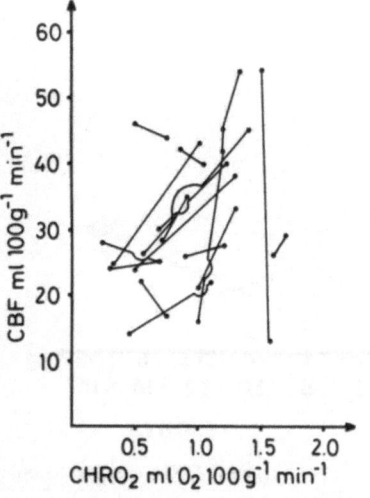

Fig. 6. The relationship between cerebral metabolic rate of oxygen ($CMRO_2$) and cerebral blood flow (CBF) in 16 patients subjected to repeated studies at an interval of days. CBF was corrected for changes in $PaCO_2$ from 34 mm Hg with 3% change in CBF per mm Hg of $PaCO_2$ change (Acta Anaesthesiol Scand 1986: 30: 453–457, with permission)

mm Hg change of $PaCO_2$ from 4.5 kPa. This principle is in accordance with Obrist *et al.* (1984).

Results: A metabolic coupling between $CMRO_2$ and CBF indicated by a positive slope of the relationship between repeated, paired studies of $CMRO_2$ and CBF was observed in 7 of 16 patients (44%) and a less pronounced coupling in two patients (13%). A negative coupling was found in five patients (31%). In two patients CBF ranged from below 15 ml to above 55 ml with very little change in $CMRO_2$. The coupling between flow and metabolism was independent of the levels of CBF measured (Fig. 6). In comparison with other studies (Obrist *et al.* 1984), no significant correlation between $CMRO_2$ and CBF was observed in patients with reduced CBF (CBF < 30 ml).

Repeated Studies of Cerebral Autoregulation (Study IV)

Material: 12 unconscious patients, Glasgow coma score < 7 at admission, median age 22 (range 14–67 years). In two patients CA was tested 4 times, in one 3 times and in nine patients twice. The patients were sedated as in study V.

As in study I CBF was calculated as initial slope index. Mean CBF was calculated as the average of the 16 regional flow values. The MABP and intraventricular pressure (IVP) were continuously recorded. CPP was defined as the difference between MABP and IVP. After a control CBF study the MABP was increased about 20–40% by a continuous angiotensin infusion. When MABP had been stabilized, the CBF measurement was repeated. In order to avoid the implication of changes in $PaCO_2$, only 3 mm Hg change in $PaCO_2$ between the two studies was tolerated. The response of the mean CBF to changes in CPP was quantified by the calcultaion of the slope of the CBF/CPP response (delta CBF/delta CPP).

Results: With the exception of two patients, repeated studies of CA indicated presence of CA during the first day after the trauma. An increase or unchanged delta CBF/delta CPP was found in the period 3 to 5 days after the injury. After 3–5 days an unchanged or decreased response was observed (Fig. 3, page 12).

In patients with a lactate concentration in ventricular fluid above up to 5 mmol/liter, the CA was usually absent. In studies where the lactate concentration was increased above this level, the CA was unimpaired "false".

Studies of Cerebral Autoregulation During Two Levels of PaCO₂ (Study VI)

Material: Eight unconscious and artificially ventilated patients (median age 29, range 16–70 years) were studied. CT scanning on admission disclosed cerebral contusion in all patients, subdural haematoma in four patients and epidural haematoma in one patient. Before and during the studies of CBF, three patients were sedated with pentobarbitone 100 mg per h and paralysed with pancuronium. Four patients were sedated with pethidine and chlorpromazine, and one with nitrous oxide 66%. In two patients, hypothermia (temperature 31 and 33 °C) was applied in order to control ICP. Before the initiation of the barbiturate coma treatment, all patients were classified as GCS < 6. CBF was studied within the first 2 days after the trauma. Three of the patients died within 11 days after the injury. Two recovered completely. One patient recovered with slight intellectual impairment and one with dementia.

rCBF was determined by the intraarterial 133-Xenon washout technique with a 8 to 16 channel Cerebrograph, and calculated as initial slope index by drawing a line which best fitted the first two minutes of the semilogarithmically displayed clearance curve. MABP and IVP were continuously recorded during the study. After a control CBF study (flow 1), the MABP was increased 20–30% by continuous angiotensin infusion. After stabilization of MABP the CBF measurement was repeated (flow 2). Before flow 3, the setting of the ventilator was changed in order to reduce PaCO₂ by about 10 mm Hg. The stabilization of PaCO₂ was checked by repeated arterial gas analysis and flow 3 was started. During flow 4 angiotensin infusion was repeated during unchanged hypocapnic level. The regional derangement of CA was quantified by the vasoreactivity impairment index (VI) (delta CBF, %/delta CPP, %), as defined by Olesen (1973). The regional CO₂ reactivity was calculated as delta CBF, %/delta PaCO₂ ,%. The statistical analyses were performed with Wilcoxons test for correlated data.

Results: In Table 1 the results of PaCO₂, MABP, ICP, and CBF are given as mean values and ranges. The levels of MABP and CPP were well preserved when flow 1 and 3 were compared. The levels of angiotensin-induced hypertension were identical in flow 2 and 4.

At high levels of PaCO₂, angiotensin infusion produced a significant increase in ICP from 12 mm to 15 mm Hg. At low levels of PaCO₂ ICP increased from 8 to 11 mm Hg. The decrease in ICP after induced hypocapnia (flow 1 to flow 3) was significant. The in-

Table 1. *Mean Values and Ranges of PaCO₂ (mmHg), MABP (mmHg), ICP (mmHg), CPP (mmHg) and Mean Hemispheric CBF (ml/100 g/min) in Eight Studies of Cerebral Autoregulation Performed During Moderate Hypocapnia (flow 1 and 2) and Pronounced Hypocapnia (flow 3 and 4) (Acta Anaesthesiol Scand 1981: 25: 397–401, with permission)*

	Flow 1	Flow 2	Flow 3	Flow 4
PaCO₂ mmHg	34.2	33.8	23.1	23.3
	29–40	27–40	18–31	19–31
MABP mmHg	72	100	73	104
	66–84	92–116	58–89	90–122
ICP mmHg	12	15	8	11
	2–24	2–33	2–13	5–22
CPP mmHg	60	85	65	93
	46–81	68–110	47–86	75–115
CBF ml	34.2	35.9	22.5	28.9
	19.5–70.9	18.8–60.3	15.5–30.1	19.5–36.5

Table 2. *Mean CBF Values in 8 Patients Where Autoregulation Was Tested During Moderate Hypocapnia (flow 1 and 2) and During Pronounced Hypocapnia (flow 3 and 4). In patients 2, 7, and 8 impaired autoregulation was only detected during pronounced hypocapnia (Acta Anaesthesiol Scand 1981: 25: 397–401, with permission)*

Patient nr.	Flow 1	Flow 2	Flow 3	Flow 4
1	28.5	34.5	15.5	19.5
2	70.9	60.3	15.9	22.3
3	23.5	32.0	18.7	27.2
4	19.9	27.4	23.9	28.0
5	32.4	36.8	24.4	30.5
6	31.9	40.5	24.8	32.2
7	19.5	18.8	26.8	34.6
8	46.6	36.7	30.1	36.5

creases in ICP levels during angiotensin infusion were identical when flow 2 and 4 were compared. The changes in ICP levels followed those of MABP. Owing to the changes in ICP values between flow 2 and flow 4 and the non-significant increase in MABP from 100 to 103 mm Hg, CPP increased from 85 to 93.

On comparing the CBF values in flow 3 and flow 4, a statistical significant flow increase was found following the induction of a MABP increase. On the other hand, no significant CBF change between flow 1 and flow 2 was observed. The changes in mean CBF in the eight studies are shown in Table 2. In patients nr 2, 7, and 8 CBF decreased during normocapnic CA tests, but increased during tests performed during hypocap-

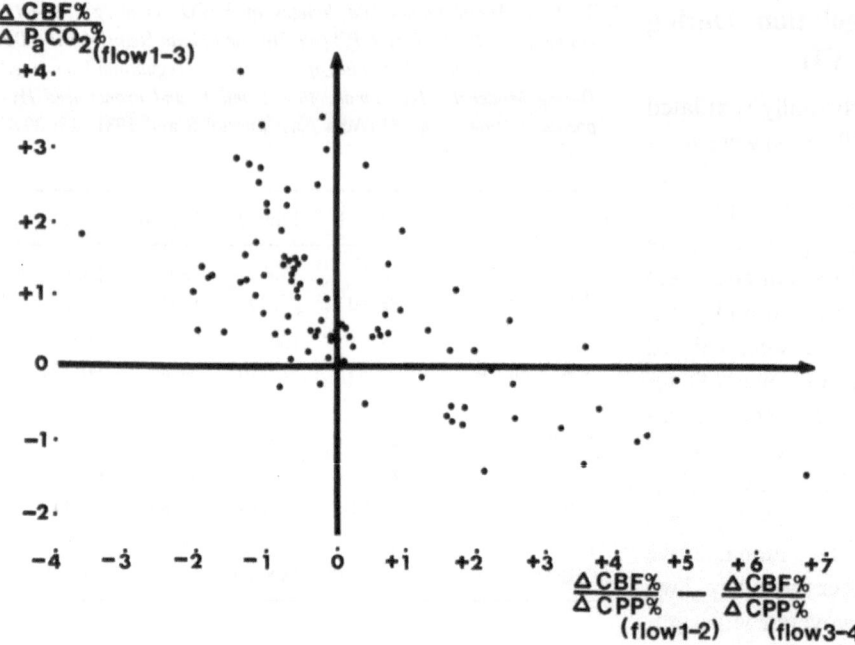

Fig. 7. The relationship between the difference in regional vasoreactivity impairment index (VI) (delta CBF%/delta PaCO$_2$%) measured during two levels of PaCO$_2$ and the regional CO$_2$ reactivity (delta CBF%/delta PaCO$_2$%) (Acta Anaesthesiol Scand 1981: 25: 397–401, with permission)

nia. The medians and ranges for the mean hemispheric vasoreactivity index changed form 0.42 (−0.58–1.04) during flow 1 and 2 to 0.80 (0.23–1.41) during flow 3 and 4.

The regional studies of CO$_2$ reactivity and VI indicated that in the majority of regions with normal CO$_2$ reactivity (positive reactivity) the VI changed from a lower to a higher value when PaCO$_2$ was reduced during hypocapnia. In comparison, in regions with negative CO$_2$ reactivity (inverse steal phenomenon), the VI was changed from a higher to a lower value in most regions. In Fig. 7 the regional changes in VI, expressed as the difference in VI (delta CBF %/delta CPP%) (flow 1–2) − (delta CBF %/delta CPP %) (flow 3–4), are correlated to the regional CO$_2$ reactivity (delta CBF %/PaCO$_2$ %) (flow 1–3).

Comments: It was not possible to state that CA improves during hypocapnia. On the contrary, CA was found to be relatively unchanged during a high level of PaCO$_2$ and impaired during induced hypocapnia. In the regional studies an improvement of CA was observed in regions with inverse steal phenomenon.

Studies of CO$_2$ Reactivity (Study I, II and VIII)

Material: 27 unconscious patients (Glasgow coma scale < 7 on admission), median age 23 years (range 9–67 years) were included. 15 studies were performed within two days of trauma, 16 studies between the third day and one week, eight studies during the second week

and six studies during the third week. In 15 patients the CO$_2$ reactivity was testet repeatedly (2–3 studies) at intervals of 2–4 days in 2 patients and one week or more in nine patients. In 21 patients angiographic studies revealed a space-occupying lesion. These patients were operated upon and cortical lesions (contusion, dilaceration, subdural haematoma) were observed during operation. In 11 of these patients neurological examination revealed impairments of the brain-stem reflexes (oculocephalic, vestibular, and pupillary reflexes), whereas these reflexes were unaffected in 10 patients.

In accordance with study V the patients were classified in two groups. Group I included 19 studies in 12 patients; Group II included 26 studies in 15 patients.

The rCBF studies were done with the intracarotid 133-Xenon washout technique with a 16 channel Cerebrograph. The CBF studies were performed over the most injured hemisphere (the site of the mass lesion). In accordance with study V the calculation of CBF was based on initial slope index and stocastic analyses. 15 minutes after the first CBF study the pulmonary ventilation was increased by 20–30%. This change resulted in a 0.5 to 1.0% decrease in end-tidal CO$_2$. When end-tidal CO$_2$ was stabilized, about 10 minutes after changing the respiratory setting, the rCBF was repeated. On the basis of the mean CBF the CO$_2$ reactivity was calculated as delta ln CBF initial/delta PaCO$_2$ mm Hg. In accordance with studies of ischaemic thresholds, rCBF in the interval $15 \leq CBF < 20$ ml/100 g/min was defined as moderate oligaemia and rCBF < 15 ml as severe oligaemia. During the study IVP and MABP were continuously recorded. Ventricular fluid was withdrawn a few hours after the CBF measurements for determination of lactate, pyruvate, pH, and bicarbonate.

Results: For the total number of observations, the hemispheric CO_2 reactivity ± SD was 0.028 ± 0.015. The values was 0.037 ± 0.015 in group I and 0.026 ± 0.012 in group II (P < 0.01, Student's test). The correlation between time after the injury and the CO_2 reactivity is shown in Fig. 8. In groups I and II a positive significant correlation was found. The steepest slope of the regression line was found in group I.

Generally the CO_2 reactivity was found to be higher in regions with tissue peaks. In Fig. 9 the regional CO_2 reactivity in 10 patients with tissue peak configuration of the clearance curves is shown. The calculation of rCBF was based on stocastic analysis. The CO_2 reactivity in tissue peak regions was found to be higher compared with regions without tissue peaks (study II). In the same study it was observed that in several regions tissue peaks disappeared during hypocapnia and the regional flow pattern became more homogeneously.

In study VIII, oligaemia (rCBF < 20 ml) was found in 5.3% of all regions before hyperventilation and in 16.1% of the regions after hyperventilation. When hyperventilation was applied the number of studies with rCBF < 20 ml increased from 10 to 21 of 45 studies. The number of patients with rCBF < 20 ml was 8 before

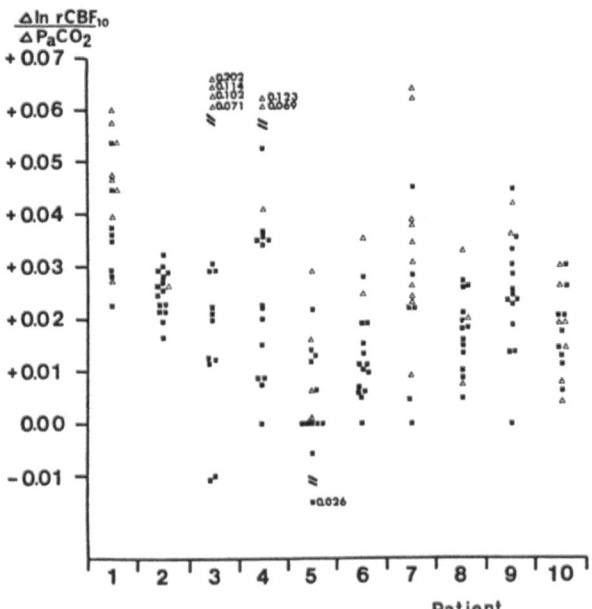

Fig. 9. The regional CO_2 reactivities measured as the proportion between delta ln rCBF/delta $PaCO_2$ mm Hg in 10 patients. rCBF was measured as stocastic flow (height over area). Open triangles: Tissue peak regions. Solid squares: Regions without tissue peaks (Acta Anaesthesiol Scand 1977: 21: 359–367, with permission)

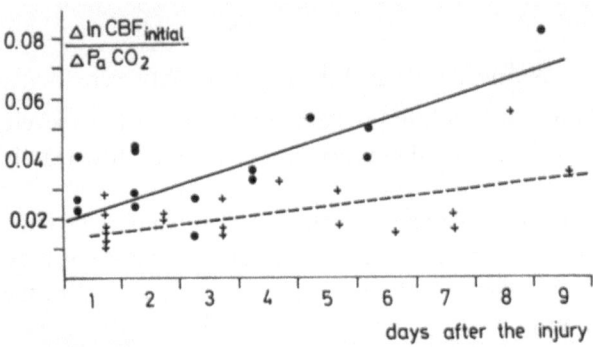

group 1: y = 0.0055x + 0.018 (r = 0.79 P < 0.01), dots: •
group 2: y = 0.0021x + 0.014 (r = 0.48 P < 0.05), stars: +

Fig. 8. The relationship between time after the injury and the CO_2 reactivity in 26 comatose patients with severe head injuries. The CO_2 reactivity was calculated as delta ln CBF/delta $PaCO_2$ mm Hg. Solid circles: Patients with good recovery or slight mental impairments (Group I). Asteriks: Patients in whom the head injury resulted in dementia, vegetative survival, or death (Group II). The regression lines are indicated (Acta Anaesthesiol Scand 1977: 21: 222–231, with permission)

and 15 after hyperventilation. The frequency of oligaemic rCBF was correlated to the outcome of the patients. In Table 3 the mean value and SD are indicated for $PaCO_2$, ICP, and CPP in patients in group I and II. Before hyperventilation the percentage of regions with moderate and severe oligaemia was 0. After hyperventilation, the percentage of regions with moderate and severe oligaemia increased to 6.3 and 1.0% respectively. In group II the percentage of regions with moderate oligaemia increased from 8.5 to 16.7% and the percentage of regions with severe oligaemia from 0 to 4.1%. In patients with a good outcome (Group I) and without regional oligaemia, mean hemispheric CBF averaged 43.2 ± 8.2 ml, contra 33.2 ± 3.9 ml in patients with regional oligaemia. In patients with poor outcome (Group II), mean hemispheric CBF averaged 40.2 ± 9.7 ml in patients without oligaemia and 25.3 ± 4.5 ml in patients with regional oligaemia. Figure 10 indicates changes in rCBF after intensified hypocapnia. Before the decrease in $PaCO_2$, rCBF values < 20 ml/100 g/min were observed in three regions. After the $PaCO_2$ decrease, regions with rCBF < 20 ml were found in 9 of 16 regions.

In patients with cortical lesions but without impairment of the brain-stem reflexes, positive correlations between the concentration of lactate and the lac-

Table 3. *The Frequency of Regions with Moderate Oligaemia Defined as (15 ≤ rCBF < 20 ml), and Severe Oligaemia (rCBF < 15 ml) Before and After Hyperventilation in Two Groups of Patients.* In group I (good recovery or slight mental impairment) 19 studies in 12 patients were performed. In group II (dementia, vegetative survival or death) 26 studies in 15 patients. The values of $PaCO_2$ (kPa), intracranial pressure (ICP) (kPa) and cerebral perfusion pressure (CPP) (kPa) before and after hyperventilation are indicated by mean values and SD. The Wilcoxon test was applied for paired data and the Mann-Whitney test for unpaired data. The Chi square test with Yates correction was used for difference between frequencies (* $P < 0.05$) (Acta Neurochir (Wien): 1989: 96: 100–106, with permission)

	Group I: Good recovery or slight mental impairment		Group II: Dementia, vegetative survival or death	
	Before hyperventilation	After hyperventilation	Before hyperventilation	After hyperventilation
$PaCO_2$ kPa	4.8 ± 1.1	3.6* ± 0.9	4.7 ± 0.7	3.5* ± 0.6
ICP kPa	2.8 ± 0.7	2.0* ± 0.5	2.5 ± 1.0	1.7* ± 0.7
CPP kPa	9.6 ± 2.0	9.6 ± 1.7	10.5 ± 2.0	10.5 ± 1.4
Number of regions with moderate oligaemia (15 ≤ rCBF < 20 ml) in % of total nr. of regions	0	6.3*	8.5	16.7*
Severe oligaemia (rCBF < 15 ml) in % of total nr. of regions	0	1.0	0	4.1*
Nr. of studies with rCBF < 20 ml	0 of 19	5 of 19	10 of 26	16 of 26
Nr. of patients with rCBF < 20 ml	0 of 12	4 of 12	8 of 15	11 of 15

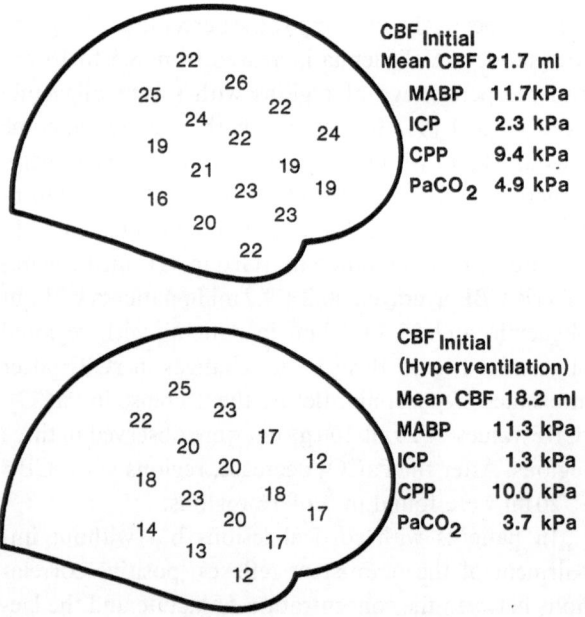

CBF Initial

Mean CBF	21.7 ml
MABP	11.7 kPa
ICP	2.3 kPa
CPP	9.4 kPa
$PaCO_2$	4.9 kPa

CBF Initial (Hyperventilation)

Mean CBF	18.2 ml
MABP	11.3 kPa
ICP	1.3 kPa
CPP	10.0 kPa
$PaCO_2$	3.7 kPa

tate/pyruvate ratio and the CO_2 reactivity were found. In patients with brain-stem lesions but without severe cortical lesions, these correlations were negative (Study I).

Comments; These studies indicate that the CO_2 reactivity is often reduced during the acute phase of head injury. The reduction in reactivity is correlated to the severity of the trauma (outcome). Tissue peak hyper-

Fig. 10. rCBF measured as initial slope index before and after hyperventilation ($PaCO_2$ change form 4.9 to 3.7 kPa) in a 19-year-old male studied on the second day after the acute head injury. The changes in mean CBF, mean arterial blood pressure (MABP), intracranial pressure (ICP), cerebral perfusion pressure (CPP), and $PaCO_2$ are indicated. Hyperventilation provoked severe oligaemia defined as rCBF < 15 ml/100 g/min in four regions (Acta Neurochir (Wien) 1989: 96: 100–106, with permission)

aemia often disappears after hyperventilation. The CO_2 reactivity is fairly high in tissue peak regions. Consequently, on hyperventilation the regional flow pattern becomes more homogeneously. In patients with low hemispheric CBF hyperventilation may produce rCBF decrease below 20 ml/100 g/min. This level is probably close to the ischaemic threshold of synaptic failure. The occurence of oligaemic flow deprivation after hyperventilation is correlated to the severity of the trauma, and low values of rCBF are observed more frequently in patients with a poor outcome.

Studies of Barbiturate- and CO₂ Reactivities (Study IX)

Material: Nine patients, median 18 years (range 12–32) with acute HI were studied with a 32-channel Cerebrograph (inhalation method, ISI based on clearance between 0.5 and 1.5 min). In five patients CBF was studied repeatedly (2–3 times) with intervals of days, in four patients CBF was only studied once. On admission all patients were unconscious with Glasgow coma score < 6. CT scanning showd diffuse swelling in one patient, subdural haematoma and contusion in 3 patients and cerebral contusion in 5 patients. The CBF measurements were done at strategic intervals either to follow the treatment (hyperventilation and barbiturate coma treatment), or to determine whether these interventions should be continued or adjusted.

rCBF was measured 2–4 times, before and after hyperventilation, and before and after injection of a thiopentone bolus (5 mg/kg i.v.). $AVDO_2$ was calculated as the difference between oxygen content in arterial and jugular venous blood. In all studies ICP and MABP were continuously recorded. $CMRO_2$ was calculated as the product of the mean CBF value of the 32 regions and $AVDO_2$. The CO_2 reactivity was calculated as relative values as delta CBF %/delta $PaCO_2$ mm Hg. The barbiturate reactivity was calculated globally as the absolute change in $CMRO_2$ (delta $CMRO_2$) and regionally as the percentage change in rCBF. (delta rCBF %).

Results: In the seven patients, 15 studies of global barbiturate reactivity were performed. In Fig. 11 the relationship between $CMRO_2$, before a bolus injection of thiopentone (5 mg/kg) and the changes in $CMRO_2$ (delta $CMRO_2$) 2 min after thiopentone injection are indicated. At high levels of $CMRO_2$ thiopentone generally resulted in a more pronounced decrease in $CMRO_2$; however, great interindividual and intraindividual differences in the changes in $CMRO_2$ were

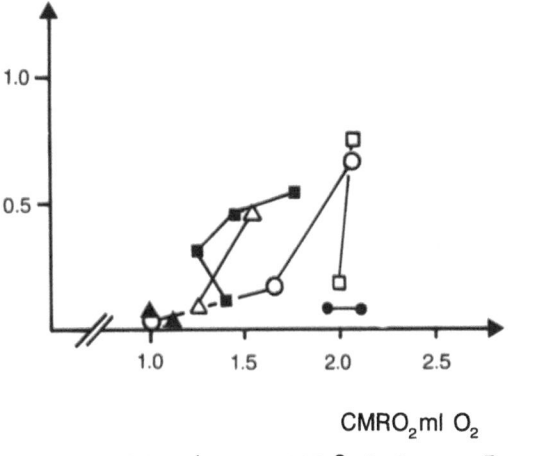

case 1 : ■—■ case 3 : △—△ case 5 : □—□
case 2 : ●—● case 4 : ○—○ other cases : ▲

Fig. 11. The relationship between cerebral metabolic rate of oxygen ($CMRO_2$) before a bolus injection of thiopentone 5 mg/kg and the change in $CMRO_2$ (delta $CMRO_2$) 2 min after thiopentone injection in seven comatose patients with acute head injury, five of whom were subjected to repeated CBF studies (Acta Neurochir (Wien) 1989: 98: 153–163, with permission)

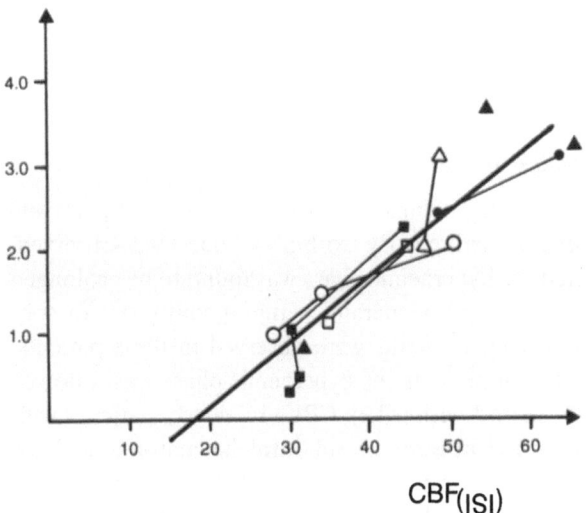

case 1 : ■—■ case 3 : △—△ case 5 : □—□
case 2 : ●—● case 4 : ○—○ other cases : ▲

Fig. 12. Indicates the correlation between CBF before hyperventilation and the absolute CO_2 reactivity (delta CBF/delta $PaCO_2$ mm Hg) in 16 studies from seven patients, five of whom were subjected to repeated studies. The correlation coefficient r = 0.9028, P < 0.01). The equation for the regression line is y = 0.07566 × − 1.378 (Acta Neurochir (Wien) 1989. 98: 153–163, with permission)

observed. At the level of $CMRO_2$ ranging from 1.1. to 1.2 ml/ O_2/100 g/min, the changes in $CMRO_2$ were found to be very small and below 0.1 ml O_2.

The absolute and relative CO_2 reactivities averaged 2.5 ml/mm Hg and 4.2% change CBF/mm Hg. The correlation between the absolute CO_2 reactivity (delta CBF/delta $PaCO_2$) and the respective values of CBF measured before $PaCO_2$ reduction are indicated in Fig. 12. Using all data a positive correlation was observed (n = 16, r = 0.9028, P < 0.01), and the equation for the regression line was y = $0.07566 \times - 1.378$. The intercept between the X-axis and the regression line was 18.2 ml. The correlation between CBF and the relative CO_2 reactivity was significant and positive (n = 16, r = 0.6370, P < 0.01). The equation for the regression line was y = $0.08067 \times + 0.7482$.

Other results concerning barbiturate reactivity:

In individual studies chronic barbiturate administration was accompanied by an abolished barbiturate reactivty. Other studies in individual patients indicated that the barbiturate reactivity was low or abolished initially in regions with severe cerebral contusion. In contrast, normalization of the barbiturate reactivity was observed during recovery (study IX).

Summary of the Nine Studies

Studies of CBF

In young victims with cortical lesions a hyperaemic phase occurring early (within 24 hours) was observed. Often the hyperaemic phase was found to be prolonged (up to 7 days). Generally, cerebral contusions or subdural haematomatas were observed in these patients. In several patients the hyperaemic phase was followed by a period with a low CBF. In elderly patients with cerebral contusion or subdural haematoma with or without abnormal brain-stem reflexes and in patients with abnormal brain-stem reflexes but without severe cortical contusion, CBF was found to be constantly low (study V).

In patients with space-occupying lesions (contusion and/or subdural haematoma or oedema) hemispheric CBF generally was highest ipselateral to the lesion (study IX).

In patients who regained conscousness an increase in CBF was generally observed (study V).

In patients with intracranial mass-lesions decompression was generally followed by an increase inCBF (study V).

In patients with reduced hemispheric CBF (CBF < 30 ml/100 g/min) the mean $AVDO_2$ was higher, and the mean value of ventricular fluid pH was lower (study VII).

Studies of $CMRO_2$

In conscious patients with severe HI $CMRO_2$ ranged from 0.24 to 2.21 ml $O_2/100$ g/min. In these severely injured patients the level of $CMRO_2$ was not related to clinical outcome or the presence of abnormal brain-stem reflexes. Repeated studies of $CMRO_2$ at intervals of days indicated that $CMRO_2$ was constantly reduced in unconscious patients. A critically low $CMRO_2$ was not defined (study III).

$CMRO_2$ was negatively correlated to ventricular fluid lactate.

In comparison with other studies (Obrist *et al.* 1984) no significant correlation between $CMRO_2$ and CBF was observed in patients with reduced CBF (CBF < 30 ml/100 g/min). On the other hand, repeated studies of CBF and $CMRO_2$ performed at intervals of days or weeks indicated a positive correlation between $CMRO_2$ and CBF in about 50% of the patients (study VII).

Studies of the CO_2 Reactivity (Global and Hemispheric Studies)

The hemispheric CO_2 reactivity was low just after the trauma, especially in patients with a poor outcome and abnormal brain-stem reflexes. In patients with a good outcome the CO_2 reactivity initially was higher, and an increase to normal values within 1–2 weeks was observed (study I).

The values of the absolute and relative global reactivities were correlated to the value of global CBF obtained before induction of hypocapnia (study IX).

The values of the absolute and relative hemispheric CO_2 reactivities were highest in patients with a high hemispheric CBF.

In patients with mainly hemispheric lesions (contusion, subdural haematoma) and normal brain-stem reflexes, a positive correlation between the relative hemispheric CO_2 reactivity and the concentration of lactate and the lactate/pyruvate ratio in the ventricular fluid was observed. In patients without evidence of severe hemispheric lesions but with abnormal brain-stem reflexes negative correlations were observed between these parametres (study I).

Studies of CO_2 Reactivity Regional Studies

A fall in $PaCO_2$ increased the homogeneity of the regional flow pattern and reduced the number of tissue peak regions. In comparison with regions without tissue peaks, the CO_2 reactivity was higher in regions with tissue peaks (study II).

With a 16 channel Cerebrograph inverse steal phenomena were observed in 11% of all regions in patients with a poor outcome, but only in 3% of the regions in patients with a good outcome. Regions with inverse steal phenomena were observed scattered over the hemispheres and generally without relation to regions with cerebral contusion. However, in a few studies the inverse steal phenomenon was observed in regions with severe cerebral contusion or haematoma (study II).

During acute hypocapnia the frequency of regions with oligaemia, defined as regions with rCBF < 20 ml/100 g/min, was increased. In patients with a poor outcome the frequency of regions with oligaemia was highest (21% of all regions), against 7% of all regions in patients with a good outcome. In patients with a reduced hemispheric CBF before hyperventilation, the frequency of regions with oligaemia after induced hypocapnia was highest (study VIII).

Adaptation to Prolonged Hyperventilation

In patients with severe HI, studies of ventricular fluid pH and bicarbonate before and after induction of prolonged artificial hyperventilation indicated that adaptation was delayed or absent (study I).

Studies of Cerebral Autoregulation (CA) (Hemispheric Studies)

During the first two days after trauma, hemispheric CA apparently was normal. On the other hand, repeated studies of hemispheric CA indicated that CA was lost 3–5 days after the trauma, and a normalization of CA occured within the first week (study IV).

In patients with a high concentration of lactate in the ventricular fluid the CA apparently was normal (Study IV).

Studies of hemispheric CA at two different levels of $PaCO_2$ indicated that during moderate hypocapnia the CA apparently was intact. However, during pronounced hypocapnia the CA generally was abolished (Study VI).

Studies of Cerebral Autoregulation (CA) (Regional Studies)

During acute increase in blood pressure the number of regions with tissue peaks increased (study IV).

In regions with severe cerebral contusion the CA generally was lost (study IV).

With a 16 channel Cerebrograph a reciprocal response to angiotensin (indicated by a 20% fall in rCBF)

was observed in 2.6% of all regions and a 10% fall in rCBF was observed in 10% of all regions. The frequency of reciprocal reactions was unrelated to changes in ICP during angiotensin infusion (study IV).

Studies of regional CA during two levels of $PaCO_2$ indicated that in regions with preserved CO_2 reactivity the CA apparently was best preserved during moderate hypocapnia. In contrast, a further reduction in $PaCO_2$ resulted in an increase in the number of regions with impaired CA. In regions with inverse steal phenomena the regional CA apparently was best preserved during pronounced hypocapnia and abolished during moderate hypocapnia (study VI).

Studies of Barbiturate Reactivity

Global barbiturate reactivity (delta $CMRO_2$ after a bolus injection of thiopentone 5 mg/kg) was dependent on $CMRO_2$ obtained before barbiturate injection. At $CMRO_2$ levels of 1.0–1.1 ml O_2/100 g/min the global barbiturate reactivity was abolished (study IX).

In individual cases, chronic barbiturate administration was accompanied by abolished barbiturate reactivity. In regions with severe cerebral contusion, studies during the first days after the trauma indicated that the regional barbiturate reactivity (delta rCBF %) was low or abolished. In contrast, during recovery a normalization of the barbiturate reactivity was observed (study IX).

Discussion of the Results

Within the last decade the dynamic changes in CBF and metabolism have been reviewed repeatedly (Enevoldsen 1980, Jennett and Teasdale 1981, Enevoldsen 1986, Sundbärg 1988). However, the implications for therapy including controlled prolonged hyperventilation, barbiturate coma treatment etc. have only sparsely been commented upon or discussed.

In patients with severe head injury neuropathological studies indicate that ischaemic damage of the brain structures is a fairly common finding (Graham et al. 1978). The ischaemic damages might be secondary to hypotensive periods (Adams et al. 1966, Adams and Graham 1976), and the changes may be caused by intracranial hypertension (Langfitt and Gennarelli 1982). The occurrence of cerebral ischaemia is supported by experimental studies indicating a correlation between the severity of the trauma on one hand, and on the other hand, the neuropathological changes, biochemical changes in cerebral tissue and ventricular fluid, with depletion of phosphocreatine and ATP, and increase in lactate (Nilsson *et al.* 1977, Tornheim *et al.* 1983, Wagner *et al.* 1985, Gennarelli *et al.* 1986, Vink *et al.* 1987). Accordingly, clinical studies in the acute phase of HI indicate an increase in ventricular fluid lactate. Other studies have shown that a persistantly high lactate level in the ventricular fluid suggests a poor outcome (Fieschi *et al.* 1974, Overgaard and Tweed 1974, Cold *et al.* 1975 b, Enevoldsen *et al.* 1976, Enevoldsen and Jensen 1978, Cold 1981, DeSalles *et al.* 1986). An increase in lactate concentration is associated with a decrease in extracellular pH (deSalles *et al.* 1987 b). Other studies indicate that ICP hypertension is a fairly common finding in the acute phase of HI (Miller *et al.* 1977). A pressure exceeding 40 mm Hg for hours is therefore associated with a poor outcome (Troupp 1967, Vapalahti *et al.* 1969, Cold *et al.* 1975, Miller *et al.* 1977, Miller *et al.* 1981, Changaris *et al.* 1987). These findings suggest that ICP hypertension is a factor of importance in the development of cerebral ischaemia. A low CPP (CPP < 60 mm Hg) for several hours has been correlated with the subsequent death of most patients. These observations suggest that ischaemic damages, owing either to ICP-hypertension or a decrease in CPP, are of major importance.

The early studies of CBF in HI revealed an inhomogeneous pattern. During the first days after the trauma CBF differed considerably. High and low CBF values were found, unrelated to the level of $PaCO_2$ and MABP (Fieschi *et al.* 1974, Overgaard *et al.* 1974, Enevoldsen *et al.* 1976). However, repeated CBF studies in 21 comatose patients showed some characteristic patterns. In relatively young patients with cerebral contusion with or without abnormal brain-stem reflexes, a hyperperfusion phase occurring within 24 hours after the trauma was observed. The duration of this phase differed considerably. In severely injured patients with impairment of the brain-stem reflexes the hyperaemic phase was followed by a period with low CBF. In patients with cerebral contusion and abnormal brain-stem reflexes another pattern was observed, with increasing CBF during the first week after the trauma of 6–8 days duration. In patients with a poor outcome and abnormal brain-stem reflexes, the hyperaemic phase was followed by period with reduced CBF. In this group of patients the ages were comparable with the ages in the group with the occurrence of hyperaemia during the first 24 hours. In elderly patients with hemispheric cortical lesions and in patients with abnormal brain-stem reflexes but without severe lesions of the cerebral hemispheres, a low perfusion phase was identified. The duration of these phases might be several

weeks (Cold and Jensen 1980). Some of these observations are identical with those observed in a study of 76 patients with HI by Overgaard (1975). This author found hemispheric ischaemia with CBF < 20 ml in 7 patients, all of whom died or remained apallic. In all patients the brain-stem reflexes were absent or impaired. In 20 patients an initial hemispheric hyperaemia was seen with CBF > 65 ml (initial slope). Regionally, tissue peaks were found and cerebral cortical contusion was observed in 13 of 20 patients. Other studies indicate that if hyperaemia is present a tendency to normalization with time occurs (Overgaard and Tweed 1974, Langfitt *et al.* 1977), whereas persistance of hyperaemia indicates a poor outcome (Fieschi *et al.* 1974). Hyperaemica has especially been seen in regions with cortical contusion. The clearance curves from these regions are characterized by an initial fast component referred to as a tissue peak. Correspondingly, early appearance of veins on arteriographic examination has been found in patients with tissue hyperaemia (Enevoldsen *et al.* 1976). Studies of local CBV by SPECT indicate an increase in local CBF in the medial margin of subdural haematoma (Kuhl *et al.* 1980). Global hyperaemia is often observed in children and young victims (Kasoff *et al.* 1972, Muizelaar *et al.* 1989 a) indicating a hyperaemic or hyperperfusion syndrom (Obrist *et al.* 1979, Cold and Jensen 1980, Obrist *et al.* 1984,). Furthermore, hyperaemia has been observed after hypotensive periods of shock and after evacuation of haematoma (decompression) (Obrist *et al.* 1984). Neuropsychological studies of outcome within the first year after HI revealed greater impairment of overall intellectual and memory functions in patients with hyperaemia in the acute phase after injury than in patients with reduced flow. In the same study patients with ICP hypertension showed greater memory deficits than did those without ICP elevations (Uzzell *et al.* 1986).

Reduced CBF has been found in elderly patients (Obrist *et al.* 1979, Cold and Jensen 1980, Obrist *et al.* 1984) and extremely low CBF values indicate impending brain death (Balslev-Jørgensen *et al.* 1972, Overgaard and Tweed 1975, Overgaard *et al.* 1981, Overgaard and Tweed 1983) and a poor prognosis (Langfitt *et al.* 1977, Overgaard and Tweed 1985, Cold 1989 a). Reduced CBF seems to indicate brain swelling caused by oedema (Overgaard and Tweed 1976, Bruce *et al.* 1973). In addition, severe oligaemia has been observed in the boundary zones between the cerebral arteries, and flow deprivation in these regions is related to a poor outcome (Overgaard and Tweed 1983). Reduced CBF has also been observed in regions adjacent to mass lesions (Bruce *et al.* 1973) and in patients with impairment of the brain-stem reflexes but without evidence of cortical lesions judged by arteriography and explorative operation. In these patients CBF might be homogenously low without indicating a poor prognosis (Cold and Jensen 1980). That CBF might be very low in patients with impaired or abolished brain-stem reflexes suggests that brain-stem lesions *per se* profoundly reduce cortical CBF and metabolism. This view is supported by reports of very low CBF and $CMRO_2$ in apallic patients and in patients with midbrain infarction (Ingvar and Sourander 1970) and by experimental studies of mesencephalic reticular formation lesions in rats where a marked reduction of rCBF and glucose utilization was observed (Hawkins *et al.* 1979). Other studies in rats with reticular formation lesions indicate that the reduction in CBF and metabolism persists throughout the four day observation period (Hass *et al.* 1976). Furthermore, in deeply comatose patients recovery of consciousness is followed by an increase or normalization of CBF (Langfitt *et al.* 1977, Obrist *et al.* 1984). In severe HI $CMRO_2$ is generally found to be low (Bruce *et al.* 1973, Cold 1978, Obrist *et al.* 1979, Obrist *et al.* 1984). A correlation between state of consciousness and $CMRO_2$ has been observed (Langfitt *et al.* 1979, Muizelaar *et al.* 1989 a). Over a large range of consciousness levels a high $CMRO_2$ indicates a good outcome (Tabaddor *et al.* 1972, Hass 1976, Jaggi *et al.* 1990). In a recent study of 96 patients with severe HI the probability of recovery was correctly predicted in 82% of the patients when age, initial GSC, and occurrence of ICP-hypertension were combined with evaluation of $CMRO_2$ (Jaggi *et al.* 1990). However, in deeply comatose patients the correlation between $CMRO_2$ and outcome is poor (Bruce and Langfitt 1976, Cold 1978, Obrist *et al.* 1984). Thus, values as low as $0.4 \, ml \, O_2/100 \, g/min$ have been found in patients who regained consciousness and recovered (Cold 1978). In another clinical study where $CMRO_2$ and CMR-lactate were measured and correlated to CT scanning, it was found that patients showing cerebral infarction at CT scanning exhibited a characteristic cerebral metabolic pattern. A $CMRO_2$ of less that $0.6 \, ml \, O_2/100 \, g/min$ in one or more studies and a markedly elevated lactate production were observed in these patients (Robertson *et al.* 1987).

The correlation between $CMRO_2$ and CBF is generally absent (Bruce *et al.* 1973, Muizelaar *et al.* 1989 a). However, Obrist *et al.* (1984) found a positive correlation in patients with reduced CBF (below 30 ml). This finding has not been observed in a recent study where repeated studies at intervals of days demonstrated a positive correlation between $CMRO_2$ and CBF in about 50% of the patients. This correlation was found to exist independently of the presence or absence of hyperaemia defined as CBF > 30 ml (Cold 1986). In the same study and in accordance with the study by Obrist *et al.* (1984), hyperaemia was associated with a high relative CO_2 reactivity.

In patients with severe HI the correlation between CBF and $AVDO_2$ is generally poor. However, in a recent study by Robertson *et al.* (1989) including 51 patients, $AVDO_2$ and AVD-lactate were studied together with global CBF, and the lactate-oxygen index ($-AVD$-lactate/$AVDO_2$) was calculated. In non-ischaemic patients, defined as lactate-oxygen index < 0.08, it was possible to define a group of hyperaemic patients with $AVDO_2 < 1.3 \, \mu mol/ml$ and a group of patients with compensated hypoperfusion where the decrease in CBF was accompanied by an increase in $AVDO_2$ and where $AVDO_2$ exceeded $3.0 \, \mu mol/ml$. In patients with ischaemia, defined as lactate-oxygen index > 0.08, the $AVDO_2$ and CBF varied considerable and generally the $CMRO_2$ was were low.

The CA has been found to be impaired in patients with severe HI (Fieschi *et al.* 1974, Enevoldsen *et al.* 1976, Cold and Jensen 1978, Enevoldsen and Jensen 1978, Muizelaar *et al.* 1989 b). Regions with cerebral contusion are especially exposed to CBF increase when CPP is augmented. Tissue peaks are provoked in these regions (Enevoldsen *et al.* 1976, Cold and Jensen 1978, Enevoldsen and Jensen 1978). In patients revovering from HI an improvement of CA has generally been observed. With a 16 channel Cerebrograph placed over the most severely injured hemisphere, regional loss of CA was observed in 83% of comatose patients. In the same study a biphasic correlation between ventricular fluid lactate and the slope of the CBF/CPP relationship was found. These findings suggest that hemispheric CA is preserved at low levels of lactate, impaired at levels of lactate ranging from 3 to 5 mmol/l, and preserved at lactate levels above 5 mmol/l (Cold and Jensen 1978). These observations together with the findings indicating intact CA during the first days after the trauma have raised the question whether intact CA is a "false" phenomenon (Fieschi *et al.* 1974). The mechanism of "false" CA is explained by an increase in tissue pressure owing to leakage of the blood-brain barrier during angiotensin infusion. In this way an increase in MABP should give rise to an increase in tissue pressure, the result being an unchanged or even reduced CPP. However, clinical studies of CA in patients with severe HI, where ICP has been monitored, seldom support this view, because generally the "false" reaction is not associated with an increase in ICP (Fieschi *et al.* 1974, Cold and Jensen 1978, Enevoldsen and Jensen 1978). That "false" CA is a fairly common phenomenon is supported by studies of CA during moderate and pronounced hypocapnia. In contradiction to clinical studies of CA in cerebral tumours and apoplexy (Paulson *et al.* 1972), a decrease in $PaCO_2$ does not improve CA in patients with HI. On the contrary, hypocapnia changes regions with intact ("false") CA to regions with impaired CA, suggesting an improvement from a state of "false" CA to impaired CA (Cold *et al.* 1981). In another study a dissociation between CA and CO_2 reactivity was found. Thus, CO_2 reactivity was impaired in patients with apparently intact CA, while the CO_2 reactivity was normal in patients with impaired CA (Enevoldsen and Jensen 1978). These observations suggest the occurrence of "false" CA as well. Whatever the reason for "false" CA, mechanisms other than the increase in ICP measured as intraventricular pressure might to be responsible. It has been proposed that intraventricular pressure does not always outline brain tissue pressure and gradients of pressures might exist within the brain. In this context, bifrontal studies of subarachoid pressure in severe head injury indicate that pressure gradients do not actually exist (Yano *et al.* 1987) and other studies of MABP, ICP, and CPP in patients with severe HI indicate that an increase in MABP is often followed by a decrease in ICP and an increase in CPP. This mechanism is supposed to be effected by autoregulatory mechanisms (Rosner 1987).

CO_2 reactivity is impaired or abolished in the acute phase of HI (Fieschi *et al.* 1974) and a low CO_2 reactivity indicates a poor outcome (Overgaard and Tweed 1974, Cold *et al.* 1977 a, Messeter *et al.* 1986). On the other hand, an improvement of CO_2 reactivity has been observed in patients who recover (Cold *et al.* 1977 a). In patients with tissue peak hyperaemia (regions with cerebral contusion), hypocapnia abolishes these compartments (Enevoldsen *et al.* 1976, Cold *et al.* 1977 b) and the CO_2 reactivity in these regions is fairly high (Cold *et al.* 1977 b). In other studies high hemispheric or global CBF is associated with a high CO_2 reactivity (Obrist *et al.* 1984, Cold 1986, Cold 1989 b). In patients with cortical lesions (contusion) a correlation between ventricular fluid lactate and the CO_2 reactivity has been observed. This finding suggests that the increase in lactate is followed by dilatation of cerebral vessels and hyperaemia. In the same study inverse steal reactions were observed in 11% of all regions (Cold *et al.* 1977 a). As previously mentioned studies of CA and CO_2 reac-

tivity indicate that "false" or inverse reactions to angiotensin-induced MABP increase are often to be found in regions with low or abolished CO_2 reactivity (Enevoldsen and Jensen 1978, Cold *et al.* 1981). In regions with very low CBF, hypocapnia might decrease CBF even further, and clinical studies suggest that moderate or severe oligaemia, with rCBF values very close to ischaemia threshold of synaptic failure, might occur (Cold 1989 a). Some studies suggest a negative correlation between $CMRO_2$ and $PaCO_2$ (Gordon and Bergvall 1977), and during hypocapnia an increase in $CMRO_2$ despite a decrease in CBF has been observed (Gennarelli *et al.* 1979). $AVDO_2$ predominantly represents extraction of global oxygen, whereas mean CBF in these studies represents cortical flow in one hemisphere. Thus, CBF in the contralateral hemisphere and in the deep structures of the brain are not involved. Under these circumstances the accuracy of the measurements is critical. In a recent study (Study IX) where the calculation of $CMRO_2$ was based in 32 detectors (16 placed over each hemisphere), hypocapnia did not change $CMRO_2$ (Cold 1989 b) and generally it is believed that $CMRO_2$ is unchanged during hypocapnia.

During prolonged controlled hyperventilation, animal studies (Raichle *et al.* 1970, Hansen *et al.* 1986, Albrecht *et al.* 1987, Muizelaar *et al.* 1988) and studies in humans with apoplectic insults (Christensen *et al.* 1974, Ellingsen *et al.* 1987) suggest adaptation of CBF, CSF-pH, and CSF-bicarbonate. However, in an uncontrolled study of patients with severe HI subjected to controlled hyperventilation for 24 hours in the subacute phase (day 4–6 after the trauma), at a time when ICP was well controlled and ventrilcular fluid lactate stabilized within 24 hours prior to application of hyperventilation, no signs of CSF-pH adaptation were found during periods ranging from 6 to 24 hours. This finding suggests that the normal mechanisms of CSF-pH adaptation are delayed or abolished in severe HI (Cold *et al.* 1977 a)

Barbiturate has been used extensively in the control of ICP hypertension. Clinical studies suggest that barbiturate coma-treatment can be used with advantage in the intensive care of patients with HI (Marshall *et al.* 1979, Saul and Ducker 1982, Hoppe *et al.* 1981). However, in both uncontrolled studies (Yano *et al.* 1986) and controlled studies, barbiturate coma-treatment does not improve recovery (Ward *et al.* 1985). Nevertheless, when ICP hypertension is present, injection of barbiturates often gives rise to a decrease ICP (Shapiro *et al.* 1974, Eisenberg *et al.* 1988). Owing to the negative inotrope effect of barbiturate MABP will also decrease and CPP will remain unchanged or even reduced.

Hench, it has been suggested that barbiturate treatment should be omitted in patients with restricted cardiac reserves. The effects of barbiturate injection on CBF and $CMRO_2$ might be reduced in the acute phase of HI. Often the CO_2 reactivity is reduced as well. In clinical studies abolished or decreased CO_2 and barbiturate reactivity indicate a poor prognosis; in contrast, preserved CO_2 reactivity is associated with an intact effect of barbiturate on CBF and $CMRO_2$ and indicates a good recovery (Messeter *et al.* 1986, Nordström *et al.* 1988). In a recent study the barbiturate reactivity, as indicated by the change in $CMRO_2$ or regionally as the change in rCBF after a bolus injection of thiopentone 5 mg/kg, was found to depend on the $CMRO_2$ value before barbiturate injection. In the same study levels of $CMRO_2$ ranging between 1.0 and 1.1 ml $O_2/100$ g/min were associated with abolished barbiturate reactivity; however, great intraindividual differences in barbiturate reactivity prohibited a general statement. In the same study the relationship between CO_2 reactivity and barbiturate reactivity was found to be poor, although highly significant correlations could be found in some patients suffering from HI (Cold 1989 b). As stated by Sawada *et al.* (1982), acute tolerance to barbiturate occurs within days of the start of therapy. Accordingly, during chronic barbiturate therapy the metabolic effect of barbiturate on cerebral metabolism is probably also diminished.

In severe HI mannitol gives rise to an increase in CBF and $CMRO_2$ (Bruce *et al.* 1973, Mendelow *et al.* 1985, Jafar *et al.* 1986). Clinical studies suggest that the increase in CBF is due to a reduction in blood viscocity (Burke *et al.* 1981, Muizelaar *et al.* 1983). Other studies suggest that the decrease in ICP is only observed in patients with an intact CA. In these patients CBF is unaltered but CVR reduced owing to a decrease in blood viscosity. In contrast, mannitol administered to patients with impaired CA will decrease blood viscosity and CVR. These changes may increase CBF while ICP remains unaltered (Muizelaar *et al.* 1984). The clinical implications of these studies are uncertain. It is reasonable to omit the use of mannitol in the control of ICP hypertension in patients with a high CBF because mannitol will increase CBF even further. On the other hand, mannitol may be used advantageously when CBF and $CMRO_2$ are low.

Summary Concerning the Dynamic Changes in Cerebral Circulation and Metabolism

Knowledge concerning the dynamic changes in CBF and metabolism and functional tests like CO_2 reactivity, CA, and barbiturate reactivity is presently overwhelming and to some extend confusing. In the following an attempt is made to elucidate the dynamic changes in cerebral circulation and metabolism after acute HI.

In children and young victims a hyperaemic phase or hyperperfusion syndrome has repeatedly been observed (Langfitt *et al.* 1977, Obrist *et al.* 1979, Muizelaar *et al.* 1989 a), and a relative high incidence of diffuse swelling is observed on CT scanning; furthermore, the mortality rate is considerably lower compared with adults (6% versus 32–52%) (Bruce *et al.* 1978). Cortical contusion and subdural haematoma might result in a global, hemispheric or focal hyperaemic phase as well. In the acute phase an uncoupling between CBF and $CMRO_2$ is observed and the ratio $CBF/CMRO_2$ is increased. Consequently, the $AVDO_2$ is low. The CA is lost. The CO_2 reactivity is normal or increased, although it might be totally abolished in the most severely injured regions. The barbiturate reactivy is inversely proportional to the depression of $CMRO_2$.

Immediately after the trauma CBF is high (hyperperfusion period), but later on CBF is normalized or reduced. The $CMRO_2$ changes in accordance with the level of consciousness. If the patient deteriorates, the

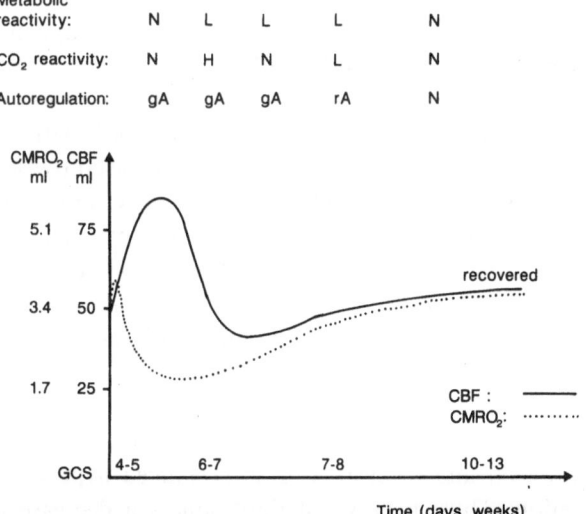

Fig. 13. The figure shows the dynamic changes in CBF and $CMRO_2$ in a patient recovering from the hyperperfusion syndrom. Changes in metabolic activity, autoregulation, CO_2 reactivity, and conscious state (Glasgow coma score (GCS)) are indicated. The metabolic reactivity and the increase in GCS follow the increase in $CMRO_2$. The CO_2 reactivity follows the CBF values. *N* indicates normal, *H* indicates high and *L* low reactivity. *gA* and *rA* indicate global and regional impairment of cerebral autoregulation. The dynamic changes are typically seen in young victims with diffuse swelling on CT scanning or in patients with cerebral contusion

CMRO$_2$ will decrease even further and CBF will decrease accordingly. This deteriotation may be caused by the development of cerebral oedema due to cerebral ischaemia or the break down of the BBB. The dynamic changes in CBF, CMRO$_2$, metabolic reactivity (barbiturate reactivity), CO$_2$ reactivity, and dynamic changes in CA are indicated in Fig. 13, where the dynamic changes may be global, or localized to one hemisphere or even smaller regions. The time axis represents days or weeks. It is postulated that different regions of the brain might be in different stages of these dynamic changes.

In adult victims the hyperaemic phase might be of shorter duration or even absent. The dynamic changes in CBF follow more or less the metabolic suppression of CMRO$_2$. At low levels of CMRO$_2$ and CBF a low metabolic reactivity to barbiturates, a low or absent CO$_2$ reactivity, and impaired or "false" CA are to be found. If the patient deteriorates, CBF and CMRO$_2$ will decrease even further. These changes are associated with suppression of consciousness. Death may occur owing to intracranial hypertension, caused by a mass lesion or cerebral oedema and the associated cerebral ischaemia. If the patient recovers, CBF, and CMRO$_2$ may increase and become normal in association with changes in consciousness. The CO$_2$ reactivity and barbiturate reactivity may improve and the CA will go through phases of lost and preserved CA. The dynamic changes are shown in Fig. 14.

In patients with diffuse lesions and impairment or absent brain-stem reflexes, and in some elderly patients, a low cerebral perfusion with a consistantly low CMRO$_2$ may occur. In these patients the CA might be "false" or impaired, and the CO$_2$ reactivity and the barbiturate reactivity are low. With improvement in the clinical condition, CBF and CMRO$_2$ may increase, the CA may normalize, and the CO$_2$ reactivity and barbiturate reactivities may improve.

The different dynamic stages may be influenced by surgical and therapeutical procedures such as decompression and medical such as prolonged hyperventilation. Thus, hyperventilation induces a decrease in CBF and an improvement of CA. On the other hand, mannitol treatment can result in an increase in CBF and CMRO$_2$. Barbiturates sedation of short duration (days?) induces a metabolic suppression with decrease in CMRO$_2$ and CBF. Chronic barbiturate sedation of week's duration may induce an increase in CMRO$_2$ and eventually abolished barbiturate reactivity.

It must be stressed that at present these considerations are speculative; however, the hyperperfusion

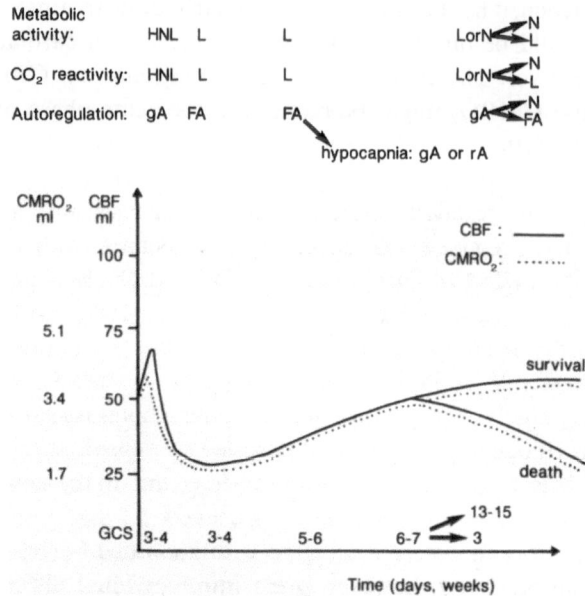

Fig. 14. The figure indicates the dynamic changes in CBF and CMRO$_2$ in elderly patients with severe head injury and patients with diffuse cerebral injury. Changes in metabolic reactivity (barbiturate reactivity), CO$_2$ reactivity, cerebral autoregulation, and conscious state (Glasgow coma score (GCS)) are indicated. The changes in patients who survived or died are indicated. The metabolic reactivity and GCS follow the CMRO$_2$, and the CO$_2$ reactivity follow the CBF values. N indicates normal, L low, and H high reactivity. gA and rA indicate global and regional impairments of cerebral autoregulation. FA indicates "false" cerebral autoregulation which eventually changes to impaired autoregulation during hypocapnia. The dynamic changes are thought to be typically seen in patients with severe cerebral contusion or subdural haematoma

syndrome as described in children and young victims is well defined, and strong evidence supports the view that elderly patients have a reduced CBF in the acute phase of HI. Other studies suggest that CO$_2$ reactivity and barbiturate reactivity are dependent on the levels of CBF and CMRO$_2$ respectively, and these reactivities are reduced in patients with a poor outcome. Furthermore, the appearance of "false" CA also indicates a poor outcome.

Therapeutic Considerations

The theory quoted above concerning the dynamic changes in CBF and metabolism has considerable therapeutic implications. Accordingly, prolonged controlled hyperventilation enhances control of ICP hypertension in stages with a high CBF. On the other hand, the ICP-reducing effect of controlled hyperventilation in phases with reduced CBF (late phase of severe HI, the acute phase of elderly victims) although present, may be of limitted therapeutic consequence. Controlled hyperventilation may be dangerous in this situation because of the decrease in MABP and CPP due the decrease in central venous filling and decrease in cardiac output; furthermore, clinical studies suggest that hyperventilation in this situation may provoke regional decreases in CBF close to ischaemic threshold.

The importance of blood pressure control can be summarized as follows: In stages with the hyperperfusion syndrome CA must be lost. Owing to the disruption of the BBB, an increase or a high level of MABP or pain might provoke cerebral oedema. On the other hand, MABP during stages of low CBF and high ICP might be marginal for cerebral perfusion, and cerebral ischaemia may threaten if MABP is reduced by therapeutic intervention. Likewise, hypovolaemia or anaemia might impede oxygen delivery to the brain.

In phases with a low $CMRO_2$ the metabolic depression induced by barbiturate may be lost or absent. Consequently, the effect of barbiturate on ICP hypertension is reduced as well. On the other hand, if $CMRO_2$ is only slightly reduced, barbiturate might reduce ICP considerably through the metabolic suppression of $CMRO_2$, CBF, and the decrease in CBV. Barbiturates depress cardiac function and reduce cardiac output and MABP and CPP. In elderly patients with reduced cardia reserves barbiturates may provoke cerebral oligaemia especially if $CMRO_2$ and CBF are reduced prior to barbiturate administration and the barbiturate reactivity is abolished.

The effects of mannitol on CBF and $CMRO_2$ are also of considerable interest. In experimental and clinical studies this drug increases flow and matabolism and decreases ICP by its osmotic effect. These combined effects might be used clinically in phases with reduced CBF and metabolism.

The theoretical effects of therapeutic intervention (hyperventilation, barbiturates and mannitol) are shown in Fig. 15 and 16. As indicated, the effect of controlled hyperventilation on CBF is most pronounced if applied during the hyperaemic phase; on the other hand, hyperventilation in phases with low flow may aggravate cerebral oligaemia. The effect of barbiturate on $CMRO_2$ is thought to depend on the resting level of metabolism, suggesting that a pronounced effect is only elicited at high levels of $CMRO_2$. In contrast, if $CMRO_2$ is low, further metabolic suppression might not occur. Under these circumstances barbiturate even might be dangerous because of failure of cerebral perfusion. If a low CBF is coupled to a reduced metabolism, mannitol may improve both cer-

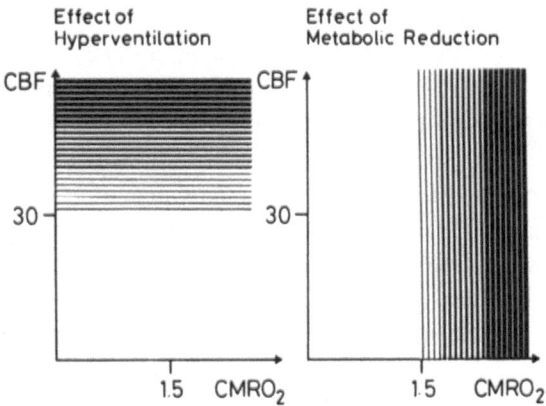

Fig. 15. The ability of hyperventilation to increase cerebral vascular resistance and reduce CBF and CBV is dependent on the CBF level. At high levels of CBF, as seen during the hyperperfusion syndrome, the CO_2 reactivity may be high, and consequently hyperventilation may increase vascular resistance and reduce CBV, CBF, and ICP. The effect of drugs, which reduce cerebral metabolism (barbiturate, etomidate etc) is most pronounced at high levels of $CMRO_2$. If on the other hand the $CMRO_2$ is below $1.5\,\mathrm{ml}\ O_2/100\,\mathrm{g/min}$, barbiturates will only reduce metabolism a very little, but may reduce cerebral perfusion pressure dangerously

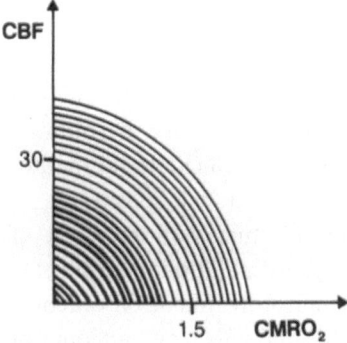

Fig. 16. Mannitol increases CBF and $CMRO_2$. These effects might be used therapeutically if CBF and/or $CMRO_2$ are low

ebral circulation and metabolism. Finally, the effects of changes in MABP and CPP might influence cerebral haemodynamics differently at different stages. If hyperperfusion is present, a high blood pressure level might aggravate the leakage of the BBB, give rise to an increase in CBV and ICP, and provoke cerebral oedema. On the other hand, if CBF initially is low a fall in blood pressure might precipitate cerebral oligaemia or even ischaemia.

Future Studies of Cerebral Circulation and Metabolism in the Acute Phase of Head Injury

Studies of CBF and metabolism are time consuming, especially if tests of cerebral autoregulation, CO_2 reactivity and barbiturate reactivity are included. Consequently, in the acute phase of head injury most studies concerning CBF and metabolism have been performed with intervals of days to weeks. The CBF studies are often performed during periods with clinical deterioration. (ie. periods with continuous high ICP or increasing ICP) and during periods with clinical improvement. In most neurosugical units the equipment for CBF measurement is stationary. Therefore patients must be transported from the intensive care unit for CBF measurement. This transport influences the ventilation regime, and sedation with hypnotic drugs and analgetics is often necessary. With transportable CBF equipment it is possible to perform daily bed-side CBF studies. Such studies might give better information because the calculated values of flow and metabolism are representative for the actual ventilator regime and sedation used in the intensive care unit.

In order to elucidate the dynamic changes in flow and metabolism in the acute phase of head injury, studies of rCBF and metabolism performed within short intervals 1–24 hours) are warranted. Ideally, these studies should be performed during controlled conditions as regards respiratory treatment and sedation, and they should include studies of CO_2 reactivity, barbiturate reactivity, and other tests which elucidate the pathophysiological changes in flow and metabolism.

Experimental and clinical studies of the effect of controlled prolonged hyperventilation, barbiturate coma-treatment, and mannitol treatment on outcome are warranted. As regards the effect on outcome of hyperventilation, barbiturate coma and mannitol treatment, experimental studies have never been performed. Clinical studies on the effect of barbiturate coma-treatment are inconclusive, and only one preliminary, controlled study concerning the effect of prophylactic artificial hyperventilation has been published. In the acute phase of HI such clinical studies should include studies of the effect on outcome and ICP hypertension of controlled hyperventilation in patients with or without a high initial CBF. Studies should also be done on the effects of barbiturate coma-treatment on clinical outcome and ICP hypertension, in patients with or without an initial high cerebral oxygen consumption, and with or without ICP hypertension, and the effect of mannitol treatment on outcome in patients with or without an initial high CBF and/or $CMRO_2$. Furthermore, experimental and clinical studies of the effects of indomethacin and Ca^{++} blocking agents on ICP, CBF, metabolism, and outcome are warranted.

With the more advanced equipment for studies of rCBF and metabolism (enhanced CT-scans, SPECT and PET), it is possible to obtain more detailed information on the regional disturbances in cerebral circulation and metabolism. These studies are warranted in order to elucidate the occurrence of high and low rCBF and metabolism, significance of steal and inverse steal phenomena, and the mechanisms of "false" cerebral autoregulation. Finally, regional changes in flow and metabolism after barbiturate, mannitol, Ca^{++} blocking agents, and indomethacin are warranted.

Summary

During the last decade several studies of cerebral blood flow (CBF) and metabolism in the acute phase of head injury have been published. It is the aim of this review to describe the dynamic changes in CBF, cerebral metabolic rate of oxygen ($CMRO_2$), cerebral autoregulation (CA), and reactivity to $PaCO_2$ and barbiturate (metabolic reactivity) in the acute phase after severe head injury and to discuss the therapeutical consequences with reference to prolonged artificial hyperventilation, hypothermia, barbiturate sedation, and mannitol therapy.

On the basis of present knowledge concerning cerebral circulation and its regulation, the author reviews the literatur concerning methodology for experimental and clinical CBF measurements and regulation of CBF and cerebral oxygen uptake. Emphasis is placed on studies of the effect of body temperatur (hypothermia) as a therapeutic tool in the control of cerebral metabolism, blood flow, and intracranial pressure. Although hypothermia significantly reduces cerebral metabolism and blood flow, the effect of hypothermia on cerebral blood flow, metabolism, ICP, and outcome after acute head injury has never been investigated in clinically controlled studies.

Experimental and clinical studies concerning sensitivity of CBF for changes in $PaCO_2$ are reviewed. The normal CO_2 reactivity defined as absolute (delta CBF/delta $PaCO_2$) and relative (%change CBF/delta $PaCO_2$) or delta ln $CBF/PaCO_2$ mm Hg are mentioned. In awake normocapnic man the relative CO_2 reactivity averages 4%/mm Hg and the absolute CO_2 reactivity 2 ml/mm Hg. Uncontrolled prospective studies show a therapeutic effect of artificially prolonged hyperventilation on outcome. Only one preliminary controlled study indicates that the outcome is poorer and recovery prolonged. Nevertheless, in the acute phase of HI, artificial hyperventilation is used rutinely for control of intracranial hypertension and during the intensive care management of the patients.

The steal and inverse steal phenomena are reviewed. Although of considerable theoretical interest these phenomena are without clinical significance in patients with head injury, unless clinical CBF measurements are performed. The frequency of the inverse steal phenomenon in studies of rCBF with a 16-channel Cerebrograph (intraarterial approach) is found to be about 10%.

During prolonged hyperventilation experimental studies and clinical studies of apoplexy show an adaptation of CBF and CSF-pH and bicarbonate. It is uncertain whether these adaptative mechanisms exist after sever head injury, because uncontrolled studies indicate that CSF-pH adaptation within 24 hours of hyperventilation is generally impaired; however, clinically controlled studies have not been done in patients with severe head injury, and clinical experience with prolonged controlled hyperventilation indicates that the effect of hyperventilation on ICP is not attenuated during treatment periods lasting several days. In the acute phase of HI, hyperventilation of 2–3 days duration generally is recommented.

In the acute phase of HI the CA is usually impaired. However, owing to the presence of "false" CA the interpretation of clinical studies of autoregulation is difficult. Clinical studies indicate that "false" autoregulation occurs fairly constantly in the acute phase after head injury. "False" CA is thought to be caused by an increase in intracranial pressure (ICP) or tissue pressure during angiotensin infusion. However, in clinical studies of autoregulation in patients with severe head injury "false" autoregulation is seldom associated with an increase in ICP corresponding to angiotensin-induced hypertension. Therefore, either local increase in tissue pressure is not transferred to ICP or other meachnisms are involved.

Experimental studies of ischaemic threshold are reviewed. The threshold for synaptic failure is found to be reached at CBF ranging from 18–23 ml/100 g/min (about 55% reduction) and threshold of membrane failure at 7–8 ml/100 g/min. In human studies failure of synaptic function has been found to be 20 ml/100 g/min. This level is reached during hypocapnia at $PaCO_2$ 2.7 kPa.

Experimental and clinical studies of the effect of barbiturate and mannitol on CBF and cerebral metabolism are reveiwed. Barbiturate induces a depression of cerebral oxygen uptake, an increase in cerebrovascular resistance (CVR), and a reduction in CBF. On the other hand, mannitol induces an increase in flow and cerebral oxygen consumption. Although barbiturates as well as mannitol treatment are able effectively to control acute intracranial hypertension, clinically controlled studies of the effect of barbiturate coma treatment are few, and the conclusions divergent. No controlled studies concerning the effect of mannitol treatment on outcome are available.

Experimental studies of head injury are reviewed with special reference to the application of impact acceleration and fluid percussion trauma. Experiments with intracranial hypertension, either with subdural or epidural balloon or by mock CSF infusion models, are also reviewed. Generally, a good correlation has been found between the severity of trauma/degree of intracranial hypertension and the pathophysiological findings and biochemical changes in cerebral metabolites (lactate, ATP, phosphocreatine, and pH). Only a few experimental studies of CBF and metabolism are available. Acute trauma is usually associated with an increase in CBF and a decrease in $CMRO_2$ but in some studies CBF and $CMRO_2$ were found to be relatively unchanged. In experimental studies the cerebral autoregulation is easily abolished in experimental studies, and the CO_2 reactivity is inversely related to the severity of the trauma.

Clinical studies of ICP and biochemical studies of CSF are reviewed. In neuropathological studies ischaemic lesions occur frequently and are thought to be caused by intracranial hypertension. ICP monitoring as a guide to therapy has been suggested in several studies. Pressure levels up to 20 mm Hg for several hours are generally accepted. However, at higher pressure levels aggressive therapy, including intensified hyperventilation and barbiturate coma treatment, is recommented.

In several clinical studies the prognostic significance of a high level of ventricular fluid lactate has been found. After acute trauma the concentration of lactate is initially high. In patients with a good outcome a decrease in lactate concentration during the first few days has been observed. In patients with poor outcome a constantly high level or an increase in lactate has repeatedly been observed.

Studies of CBF and metabolism in the acute phase of head injury are reviewed. The author's material includes 57 patients studied in three periods between 1970 and 1986. On admission all patients were deeply comatose (Glasgow coma score < 7). In 48 patients rCBF was measured by the intraarterial Xenon-133 washout technique, in 9 patients by the inhalation technique.

The median age was 23 years (range 7–70 years). In 40 patients the CBF studies were performed before CT-scanning was available. In these patients the diagnosis were based on arteriography and operative findings. In the remaining 17 patients CT-scanning was performed on admission and repeated at intervals of days or weeks. In 27 patients the CO_2 reactivity was tested, in 18 patients the cerebral autoregulation. Studies of $CMRO_2$ were performed in 22 patients. In eight patients cerebral autoregulation was tested during two levels of $PaCO_2$; in nine patients measurements of $CMRO_2$, CO_2 reactivity (8 patients), and barbiturate reactivity (7 patients) were performed. In 55 patients intracranial pressure was continuously monitored. All patients were given respirator treatment. Repeated neurological examination including the state of consciousness and evaluation of the brain-stem reflexes (pupillary, oculocephalic, vestibular reflexes to cold wear) were performed. The diagnosis based on arteriography and operative findings (40 Patients) and CT-scanning (17 patients) were as follows: Cerebral contusion or subdural haematoma without impairment of the brain-stem reflexes (15 patients), cerebral contusion or subdural haematoma with impairments of the brain-stem reflexes (35 patients), diffuse cerebral injury with impairment of the brain-stem reflexes (6 patients), diffuse cerebral swelling without impairment of the brain-stem reflexes (1 patient). An outcome study was performed from 6 months to 2 years after the trauma. The first 40 patients were classified in 2 groups as follows: Group 1, complete recovery or slight mental impairment without dementia (the patients were able to resume their previous work or school) 22 patients. Group 2, dementia present (7 patients) (vegetative survival and patients who died without gaining consciousness) 11 patients. In the remaining 17 patients outcome studies were performed in accordance with the method of Teasdale and Jennett (1976).

Studies of repeated CBF show three patterns of dynamic changes. In children and young victims a hyperaemic or hyperperfusion phase often is present immediately after the trauma. This pattern is associated with a decrease in $CMRO_2$, loss of autoregulation, and preserved CO_2 reactivity. In patients with cerebral contusion, dilaceration and in patients with cerebral haematoma a prolonged regional, hemispheric or global hyperaemia occurring after 1–2 days and lasting several days to weeks is observed. In these patients $CMRO_2$ is constantly reduced. Initially, impaired or "false" cerebral autoregulation, impaired CO_2 and barbiturate reactivities and a low $CMRO_2$ are generally observed;

later these changes are followed by normalization or a high CO_2 reactivity. With time (days or weeks) the hyperaemic phase is followed by a phase with reduced CBF and low or normal CO_2 reactivity. A persistantly low cerebral perfusion is observed in patients with diffuse brain lesions and impairment of the brain-stem reflexes. This pattern is thought to be associated with a low $CMRO_2$, low barbiturate, and CO_2 reactivities, and the presence of "false" autoregulation. The level of consciousness follows the degree of $CMRO_2$ depression. The barbiturate reactivity (defined as change in $CMRO_2$ after a bolus dosis thiopentone 5 mg/kg) correlates with the level of $CMRO_2$. The absolute and relative CO_2 reactivities are correlated to the level of CBF. Thus, the present study suggests that a high CBF is correlated to a high CO_2 reactivity, and a high $CMRO_2$ is correlated to a high barbiturate reactivity. Studies of rCBF before and after application of hypocapnia indicate that hypocapnia, in regions with relatively low rCBF, may provoke severe flow deprivation defined as rCBF < 20 ml/100 g/min. Flow deprivation after hyperventilation is especially observed in patients with low hemispheric flow and flow deprivation is associated with a poor outcome. Furthermore, a low CO_2 reactivity is associated with a poor outcome, and other studies indicate that "false" cerebral autoregulation and low reactivity to barbiturate are of prognostic significance, and indicate a poor outcome.

On the basis of the dynamic changes of CBF, $CMRO_2$, CO_2 reactivity and barbiturate reactivity in the acute phase of head injury, principles of treatment with prolonged hyperventilation, barbiturate coma treatment and mannitol treatment are suggested in the care of patients after neuroradiological investigation and neurosurgical exploration.

Hyperventilation is suggested as a rationel treatment in patients with hyperaemia with preserved CO_2 reactivity. On the other hand hyperventilation should be avoided or restricted in phases with low CBF, because hypocapnia in this situation might provoke severe oligaemia and flow deprivation close to or below threshold of synaptic failure (< 20 ml).

The effect of barbiturate on cerebral metabolism is dependent on the suppression of $CMRO_2$. In patients with a relatively high $CMRO_2$, barbiturate is able to decrease the oxygen consumption further; however, in patients with a low oxygen consumption barbiturate is unable to decrease oxygen consumption. Thus, barbiturate treatment is only indicated in patients with a high $CMRO_2$, and hyperventilation only when CBF and CO_2 reactivity are high. If $CMRO_2$ is very low (about 1.1 ml O_2/100 g/min) the barbiturate reactivity may be abolished. Under these circumstances barbiturate might decrease systemic blood pressure and CPP and may jeopardize cerebral circulation. Accordingly, if rCBF is low before the use of hypocapnia hyperventilation may reduce systemic blood pressure and CPP and provoke cerebral oligaemia.

Experimental and clinical studies indicate that mannitol increases CBF and $CMRO_2$. It is suggested that mannitol treatment should be restricted to circumstances where intracranial hypertension is associated with reduction of CBF and $CMRO_2$.

It is concluded that clinical studies of CBF, $CMRO_2$, CO_2 reactivity and barbiturate reactivity might be of help in the management of patients with severe head injury, especially if or when decisions concerning prolonged hyperventilation, barbiturate coma-treatment, and mannitol treatment are required.

References

Abdel-Dayem HM, Sadek SA, Kouris K, Bahar RH, Higazi I, Eriksson S, Englesson SH, Berntman L, Sigurdsson GH, Foad M, Olivecrona H (1987) Changes in cerebral perfusion after acute head injury: comparison of CT with Tc-99m HM-PAO SPECT. Radiology 165: 221–226

Abou-Madi MN, Trop D, Villemure JG (1983) Effect of changing PaCO₂ on intracranial pressure response to bolus infusion of mannitol. Anesthesiology 59: A 391

Abou-Madi M, Trop D, Abou-Madi N, Ravussin P (1987) Does a bolus of mannitol initially aggravate intracranial hypertension? Br J Anaesth 59: 630–639

Adams JE, Elliot H, Sutherland VC, Wylie EJ, Dunbar RD (1957) Cerebral metabolic studies of hypothermia in the human. Surg Forum 7: 535–539

Adams JH, Brierley JB, Connor RCR, Treip CS (1966) The effects of systemic hypotension upon the human brain. Clinical and neuropathological observations in 11 cases. Brain 89: 235–268

– Graham DI (1976) The relationship between ventricular fluid pressure and the neuropathology of raised intracranial pressure. Neuropathol Appl Neurobiol 2: 323–332

Agnoli A, Battistini N, Bozzao L, Fieschi C (1965) Drug action on regional cerebral blood flow in cases of acute cerebrovascular involvment. Acta Neurol Scand 41 [Suppl] 14: 142–144

– Principe M, Priori AM, Bozzoa I, Fieschi C (1969) Measurements of rCBF by intravenous injection of 133-Xenon. In: Brock M, Fieschi C, Ingvar DH, Lassen NA, Schurmann K (eds) Cerebral blood flow Springer Berlin Heidelberg New York, pp 31–34

Aitken PG, Schiff SJ (1986) Barbiturate protection against hypoxic neuronal damage in vitro. J Neurosurg 65: 230–232

Alavi A, Weller S, Alves W, Spielman G, Reivich M, Generelli T (1989) Distinct patterns of metabolic changes in head injury as detected by positron emission tomography. J Cereb Blood Flow Metab 9 [suppl] 1: S 379

Alberico AM, Ward JD, Choi SC, Marmarou A, Young HF (1987) Outcome after severe head injury. Relationship to mass lesions, diffuse injury, and ICP course in pediatric and adult patients. J Neurosurg 67: 648–656

Albrecht RF, Miletich DJ, Rosenberg R, Zahep B (1977) Cerebral blood flow and metabolic changes from induction to onset of anesthesia with halothane or pentobarbital. Anesthesiology 47: 252–256

– – Ruttle M (1987) Cerebral effect of extended hyperventilation in unanesthetized goats. Stroke 18: 649–655

Alexander SC, Smith TC, Strobel G, Stephen GW, Wollman H (1968) Cerebral carbohydrate metabolism of man during respiratory and metabolic alkalosis. J Appl Physiol 24: 66–72

Allen GD, Morris LE (1962) Central nervous system effects of hyperventilation during anaesthesia. Br J Anaesth 24: 296–306

Allen SJ, Giezentanner A, Cronau LH, Bull JM, Laine GA (1986) Whole body hyperthermia increases cerebral blood flow and impairs cerebral autoregulation. Anesthesiology 65: A 321

Altenburg BM, Michenfelder JD, Theye RA (1969) Acute tolerance to thiopental in canine cerebral oxygen consumption studies. Anesthesiology 31: 443–448

Andersen BJ, Unterberg AW, Clarke GD, Marmarou A (1988) Effect of posttraumatic hypoventilation on cerebral energy metabolism. J Neurosurg 68: 601–607

Andersen RE, Sundt TM (1983) Brain pH in focal cerebral ischaemia and the protective effects of barbiturate anesthesia. J Cereb Blood Flow Metab 3: 493–497

Arnfred I, Secher O (1962) Anoxia and barbiturates. Arch Int Pharmacodyn Ther 139: 67–74

Artru AA, Michenfelder JD (1980) Effects of hypercarbia on canine cerebral metabolism and blood flow with simultaneous direct and indirect measurement of blood flow. Anesthesiology 52: 466–469

– – (1981) Influence of hypothermia or hyperthermia alone or in combination with pentobarbital or phenotoin on survival time in hypoxic mice. Anesth Analg 60: 867–870

– Colley PS (1984) Cerebral blood flow responses to hypocapnia during hypotension. Stroke 15: 878–883

– (1987) Reduction of cerebrospinal fluid pressure by hypocapnia: Changes in cerebral blood volume, cerebrospinal fluid volume, and brain tissue water and electrolytes. J Cereb Blood Flow Metab 7: 471–479

– Hornbein TF (1987) Prolonged hypocapnia does not alter the rate of CSF production in dogs during halothane anesthesia or sedation with nitrous oxide. Anesthesiology 67: 66–71

– Katz RA, Colley PS (1989) Autoregulation of cerebral blood flow during normocapnia and hypocapnia in dogs. Anesthesiology 70: 288–292

Astrup J, Symon L, Branston NM, Lassen NA (1977) Cortical evoked potential and extracellular K⁺ and H⁺ at critical levels of brain ischaemia. Stroke 8: 51–57

– Blennow G, Nilsson B (1979) Effects of reduced cerebral blood flow upon EEG pattern, cerebral extracellular potassium, and energy metabolism in the rat cortex during bicuculline-induced seizures. Brain Res 177: 115–126

– Skovsted P, Gjerris F, Sørensen HR (1981 a) Increase in extracellular potassium in the brain during circulatory arrest: Effects of hypothermia, lidocaine, and thiopental. Anesthesiology 55: 256–262

– Møller Sørensen P, Sørensen HR (1981 b) Inhibition of cerebral oxygen and glucose consumption in the dog by hypothermia, pentobarbital, and lidocaine. Anesthesiology 55: 263–268

– (1982) Energy-requiring cell functions in the ischemic brain. Their critical supply and possible inhibition in protective therapy. J Neurosurg 56: 482–497

Auer L (1978) Origin and localization of Evan's blue extravasation in acutely-induced hypertension in cats. Eur Neurol 17: 211–215

Auer LM, Haselsberger K (1987) Effect of intravenous mannitol on cat pial arteries and veins during normal and elevated intracranial pressure. Neurosurgery 21: 142–146

Aukland K, Bower BF, Berliner RW (1964) Measurement of local blood flow with hydrogen gas. Circ Res 14: 164–187

Austin G, Horn N, Rouhe S, Hayward W (1972) Description and early results of an intravenous radioisotope technique for measuring regional cerebral blood flow in man. Europ Neurol 8: 43–51

Bakay L, Lee JC, Lee GC, Peng J (1977) Experimental cerebral concussion. Part 1: An electron microscopic study. J Neurosurg 47: 525–531

Bakay RAE, Ward AA Jr. (1983) Enzymatic changes in serum and cerebrospinal fluid in neurological injury. J Neurosurg 58: 27–37

Balslev-Jørgensen P, Heilbrun MP, Boysen G, Rosenklint A, Jørgensen ED (1972) Cerebral perfusion pressure correlated with regional cerebral blood flow, EEG and aortocervical arteriography in patients with severe brain disorders progressing to brain death. Europ Neurol 8: 207–212

Battistini N, Casacchia M, Bartolini A, Bava G, Fieschi C (1969) Effects of hyperventilation on focal brain damage following middle cerebral artery occlusion. In: Brock M, Fieschi C, Ingvar DH, Lassen NA, Schurmann K (eds) Cerebral blood flow, clinical and experimental results. Springer, Berlin Heidelberg New York, pp 249–253

Baughman VL, Hoffman WE, Miletich DJ, Albrecht RF (1986) Effects of phenobarbital on cerebral blood flow and metabolism in young and aged rats. Anesthesiology 65: 500–505

Baumbach GL, Heistad DD (1985) Heterogeneity of brain blood flow and permeability during acute hypertension. Am J Physiol 249: H 629–637

Becker DP, Miller JD, Ward JD, Greenberg RP, Young HF, Sakalas R (1977) The outcome from severe head injury with early diagnosis and intensive management. J Neurosug 47: 491–502

– – Sweet RC, Young HF, Sullivan H, Griffith RL (1979) Head injury management. In: Popp AJ (ed) Neural trauma. Raven Press, New York, pp 313–328

– (1983) Selecting patients for intracranial pressure monitoring in severe head injury. In: Ishii S, Nagai H, Brock M (eds) Intracranial pressure V. Springer, Berlin Heidelberg New York, pp 512–516

Bell BA, Symon L, Branston NM (1985) CBF and time thresholds for the formation of ischemic cerebral edema, and effect of reperfusion in baboons. J Neurosurg 62: 31–41

Beresford HR, Posner JB, Plum F (1969) Changes in brain lactate during induced cerebral seizures. Arch Neurol 20: 243–248

Bering EA (1961) Effect of body temperature change on cerebral oxygen consumption of the intact monkey. Am J Physiol 200: 417–419

– (1974) Effects of profound hypothermia and circulatory arrest on cerebral oxygen metabolism and cerebrospinal fluid electrolyte composition in dogs. J Neurosurg 39: 199–205

Berntman L, Welsh FA, Harp JR (1981) Cerebral protective effect of low-grade hypothermia. Anesthesiology 55: 495–498

Betz E, Heuser D (1967) Cerebral cortical blood flow during changes of acid-base equilibrium of the brain. J Appl Physiol 23: 726–733

Bill A, Linder J, Linder M (1976) Sympathetic control of cerebral blood flow in acute arterial hypertension. Acta Physiol Scand 96: 114–121

Bleyaert AL, Nemoto EM, Safar P, Stezoski SW, Mickell JL, Moossy J, Rao GR (1978) Thiopental amelioration of brain damage after global ischemia in monkeys. Anesthesiology 49: 390–398

Blomqvist P, Wieloch T (1985) Ischaemic brain damage in rats following cardiac arrest using a long-term recovery model. J Cereb Blood Flow Metab 5: 420–431

Boels PJ, Verbeuren TJ, Vanhoutte PM (1985) Moderate cooling depresses the accumulation and the release of newly synthetized catecholamines in isolated canine saphenous veins. Experientia 41: 1374–1377

Bolwig TG, Qistorff B (1973) In vivo concentration of lactate in the brain of conscious rats before and during seizures: a new ultra-rapid technique for freeze-sampling of brain tissue. J Neurochem 21: 1345–1348

Bouzarth WF, Kazi KH, Bubelis I, Shenkin HA (1967) Effect of temperature upon craniocerebral trauma. JAMA 199: 567–569

Boyd RJ, Connolly JE (1961) Tolerance of anoxia of the dog's brain at various temperature. Surg Forum 12: 408–410

Bozza MM, Maspes PE, Rossanda M (1961) The control of brain volume and tension during intracranial operations. Br J Anaesth 33: 132–147

Branston NM, Symon L, Crockard HA, Pasztor E (1974) Relationship between the cortical evoked potential and local cortical blood flow following acute middle cerebral artery occlusion in the baboon. Exp Neurol 45: 195–208

– Strong AJ, Symon L (1977) Extracellular potassium activity, evoked potential and tissue blood flow: Relationship during progressive ischaemia in baboon cerebral cortex. J Neurol Sci 32: 305–321

– Hope DT, Symon L (1979) Barbiturates in focal ischaemia of primate cortex: effects on blood flow distribution, evoked potential and extracellular potassium. Stroke 10: 647–653

– Ladds A, Symon L, Wang AD (1984) Comparison of the effects of ischaemia on early components of the somatosensory evoked potential in the brainstem, thalamus and cerebral cortex. J Cereb Blood Flow Metab 4: 68–81

Brawley BW, Strandness DE, Kelly WA (1967) The physiologic response to therapy in experimental cerebral ischaemia. Arch Neurol 17: 180–187

Bricolo AP, Glick RP (1981) Barbiturate effects on acute experimental intracranial hypertension. J Neurosurg 55: 397–406

Brierley JB, Graham DI (1984) Hypoxia and vascular disorders of the central nervous system. In: Hume Adams J, Corsellis JAN, Duchen LW (eds) Greenfields neuropathology. Edwards Arnold, London, pp 125–207

Brodersen P, Paulson OB, Bolwig TG, Rogon ZE, Rafaelsen OJ, Lassen NA (1973) Cerebral hyperemia in electrically induced epileptic seizures. Arch Neurol 28: 334–338

Brown SC, Lam AM, Manninen PH (1986) Haemodynamic effects of high-dose mannitol in man. Canad Anaesth Soc J 33: S 92–S 93

Bruce DA, Langfitt TW, Miller JD, Schutz H, Vapalahti M, Stanek A, Goldberg HI (1973) Regional cerebral blood flow, intracranial pressure, and brain metabolism in comotose patients. J Neurosurg 38: 131–145

– – (1976) The prognostic value of ICP, CPP, CBF, and $CMRO_2$ in head injury. In: McLaurin RL (ed) Head injuries. Grune and Stratton, New York, pp 23–25

– Schut L, Bruno LA, Wood JH, Sutton LN (1978) Outcome following severe head injuries in children. J Neurosurg 48: 679–688

Burke AM, Quest DO, Chien S, Cerri C (1981) The effects of mannitol on blood viscosity. J Neurosurg 55: 550–553

– Greenberg JH, Sladky J, Reivich M (1987) Regional variation in cerebral perfusion during acute hypertension. Neurology 37: 94–99

Busto R, Dietrich WD, Globus MY, Valdes I, Scheinberg P, Ginsberg MD (1987) Small differences in intraischaemic brain temperature critically determine the extend of ischaemic neuronal injury. J Cereb Blood Flow Metab 7: 729–738

Cain SM (1963) An attempt to demonstrate cerebral anoxia during hyperventilation of anesthetized dogs. Am J Physiol 204: 323–326

Carlsson C, Harp JR, Siesjö BK (1975) Metabolic changes in the cerebral cortex of the rat induced by intravenous pentothalsodium. Acta Anaesthesiol Scand [Suppl] 57: 7–17

– Hägerdal M, Siesjö BK (1976 a) Protective effect of hypothermia in cerebral oxygen deficiency caused by arterial hypoxia. Anesthesiology 44: 27–34

– – – (1976 b) The effects of hyperthermia upon oxygen consumption and upon organic phosphate, glycolytic metabolites, citric and cycle intermediates and associated amino acids in rat cerebral cortex. J Neurochem 26: 1001–1006

Changaris DG, McGraw CP, Richardson JD, Garretson HD, Arpin EJ, Shield CB (1987) Correlation of cerebral perfusion pressure and glasgow coma scale to outcome. J Trauma 27: 1007–1012

Chapman AG, Nordström C-H, Siesjö BK (1978) Influence of phenobarbital anesthesia on carbohydrate and amino-acid metabolism in rat brain. Anesthesiology 48: 175–182

Christensen MS (1974) Acid-base changes in cerebrospinal fluid and blood, and blood volume changes following prolonged hyperventilation in man. Br J Anaesth 46: 348–357

– (1976) Prolonged arteficial hyperventilation in cerebral apoplexy. Acta Anaesthesiol Scand [Suppl] 62: 1–24

Clifton GL, Ziegler MG, Grossmann RG (1981) Circulating catecholamines and sympathetic activity after head injury. Neurosurgery 8: 10–14

– Robertson CS, Kyper K, Taylor AA, Dhekne RD, Grossman RG (1983) Cardiovascular response ro severe head injury. J Neurosurg 59: 447–454

Cohen PJ, Wollman H, Alexander SC, Chase PE, Behar MG (1964) Cerebral carbohydrate metabolism in man during halothane anaesthesia. Effects of PaCO$_2$ on some aspects of carbohydrate utilization. Anesthesiology 25: 185–191

– (1981) To dream the impossible dream. (editorial view). Anesthesiology 55: 491–493

Cold GE, Enevoldsen E, Malmros R (1975 a) The prognostic value of continuous intraventricular pressure recording in unconscious brain-injury patients under controlled ventilation. In: Lundberg N, Ponten U, Brock M (eds) Intracranial pressure II. Springer, Berlin Heidelberg New York, pp 517–521

– – – (1975 b) Ventricular fluid lactate, pyruvate, bicarbonate and pH in unconscious patients subjected to controlled ventilation. Acta Neurol Scand 52: 187–195

– Jensen FT, Malmros R (1977 a) The cerebrovascular CO$_2$ reactivity during the acute phase of brain injury. Acta Anaesthesiol Scand 21: 222–231

– – – (1977 b) The effects of PaCO$_2$ reduction on regional cerebral blood flow in the acute phase of brain injury. Acta Anaesthesiol Scand 21: 359–367

– (1978) Cerebral metabolic rate of oxygen (CMRO$_2$) in the acute phase of brain injury. Acta Anaesthesiol Scand 22: 249–256

– Jensen FT (1978) Cerebral autoregulation in unconscious patients with brain injury. Acta Anaesthesiol Scand 22: 270–280

– – (1980) Cerebral blood flow in the acute phase after head injury. Part 1: Correlation to age of the patients, clinical outcome and localization of the injured region. Acta Anaesthesiol Scand 24: 245–251

– (1981) Cerebral blood flow in the acute phase after head injury. Part 2: Correlation to intraventricular pressure (IVP), cerebral perfusion pressure (CPP), PaCO$_2$, ventricular fluid lactate, lactate/pyruvate ratio and pH. Acta Anaesthesiol Scand 25: 332–335

– Christensen MS, Schmidt K (1981) Effect of two levels of induced hypocapnia on cerebral autoregulation in the acute phase of head injury coma. Acta Anaesthesiol Scand 25: 397–401

– (1986) The relationship between cerebral metabolic rate of oxygen and cerebral blood flow in the acute phase of head injury. Acta Anaesthesiol Scand 30: 453–457

– (1989 a) Does acute hyperventilation provoke severe cerebral oligaemia in comatose patients after head injury? Acta Neurochir (Wien) 96: 100–106

– (1989 b) Measurements of CO$_2$ reactivity and barbiturate reactivity in patients with severe head injury. Acta Neurochir (Wien) 98: 153–163

Collins RC, Posner JB, Plum F (1970) Cerebral energy metabolism during electroshock seizures in mice. Am J Physiol 218: 943–950

Connolly JE, Boyd RJ, Calvin JW (1962) The protective effect of hypothermia in cerebral ischemia: Experimental and clinical applicaion by selective brain cooling in the human. Surgery 52: 15–24

Corkill G, Chikovanni OI, McLeish I, Mc Donald LW, Youmans JR (1976) Timing of pentobarbital administration for brain protection in experimental stroke. Surg Neurol 5: 147–149

– Sivalingham S, Reital JA, Gilroy BA, Helphrey MG (1978) Dose dependency of the post-insult protective effect of pentobarbital in the canine experimental stroke model. Stroke 9: 10–12

Cote CJ, Greenhow E, Marshall BE (1979) The hypotensive response to rapid intravenous administration of hypertonic solutions in man and in the rabbit. Anesthesiology 50: 30–35

Cotev S, Severinghaus JW (1969) Role of cerebrospinal fluid pH in management of respiratory problems. Anesth Analg 48: 42–47

Crockard HA, Taylor AR (1972) Serial CSF lactate/pyruvate values as a guide to prognosis in head injury coma. Europ Neurol 8: 151–157

– Gadian DG, Frackowiak RSJ, Proctor E, Allen K, Williams SR, Russell RWR (1987) Acute cerebral ischaemia: Concurrent changes in cerebral blood flow, energy metabolites, pH and lactate measured with hydrogen clearance and 31 P and 1 H nuclear magnetic resonance spectroscopy. II. Changes during ischaemia. J Cereb Blood Flow Metab 7: 394–402

Dahlgren N, Nilsson B, Sakabe T, Siesjö BK (1981) The effect of indomethacin on cerebral blood flow and oxygen consumption in the rat at normal and increased carbon dioxide tensions. Acta Physiol Scand 111: 475–485

Daw EF, Moffitt EA, Michenfelder JD, Terry HR (1964) Profound hypothermia. Can Anesth Soc J 11: 382–393

Darby JM, Yonas H, Marion DW, Latchaw RE (1988) Local "inverse steal" induced by hyperventilation in head injury. Neurosurgery 23: 84–88

Deligné P, David M (1966) Hibernation artificielle en neuro-chirurgie. Evolution de nos techniques. Ann Anesth Franc 7: 117–129

DeSalles AAF, Kontos HA, Becker DP, Yang MS, Ward JD, Moulton R, Gruemer HD, Lutz H, Maset AL, Jenkins L, Marmarou A, Muizelaar P (1986) Prognostic significance of ventricular CSF lactic acidosis in severe head injury. J Neurosurg 65: 615–624

– Muizelaar P, Young HF (1987 a) Hyperglycemia, cerebrospinal fluid lactic acidosis, and cerebral blood flow in severely head-injured patients. Neurosurgery 21: 45–50

– Kontos HA, Ward JD, Marmarou A, Becker DP (1987 b) Brain tissue pH in severely head-injured patients: A report of three cases. Neurosurgery 20: 297–301

DeWitt DS, Jenkins LW, Lutz H et al (1981) Regional cerebral blood flow following fluid percussion injury. J Cereb Blood Flow Metab [Suppl] 1: 579–580

– – Wei EP, Lutz H, Becker DP, Kontos HA (1986) Effects of fluid-percussion brain injury on regional cerebral blood flow and pial arteriolar diameter. J Neurosurg 64: 787–794

Dhawan V, Haughton VM, Thaler HT, Lu HC, Rottenberg DA (1984) Accuracy of stable xenon/CT measurements of regional cerebral blood flow: Effect of extrapolated estimates of brain-blood partition coefficients. J Comput Assist Tomogr 8: 208–212

Diemer NH, Siemkowicz E (1981) Regional neuron damage after cerebral ischemia in the normo- and hypoglycemic rat. J Neuropath Exp Neurol 7: 217–227

Domenech RJ, Hoffman JIE, Noble MIN, Saunders KB, Henson JR, Subijanto S (1969) Total and regional coronary blood flow measured by radioactive microspheres in conscious and anesthetized dogs. Circ Res 25: 581–596

Donley RF, Sundt TM, Anderson RE, Sharbrough FW (1975) Blood flow measurements and the "look through" artifact in focal ischemia. Stroke 6: 121–131

Drayer BP, Wolfson SK Jr, Reinmuth OM, Dujovny M, Boehnke M, Cooke EE (1978) Xenon enhanced computed tomography for the analysis of cerebral integrity, perfusion and blood flow. Stroke 9: 123–212

– Gur D, Yonas H, Wolfson SK Jr, Cook EE (1980) Abnormality of the xenon brain-blood partition coefficient and blood flow in cerebral infarction: An in vivo assessment using transmission computed tomography. Radiology 135: 349–354

DuCailar J, Rioux J, Groleau D, Villard F (1964) Hypothermie au dessous de 25 par refrigeration externe et sans circulation extracorporelle. Ann Anesth Franc 4: 781–800

Duckrow RB, LaManna JC, Rosenthal M, Levasseur JE (1981) Oxydative metabolic activity of the cerebral cortex after fluid-percussion head injury in the cat. J Neurosurg 54: 607–614

Dutka AJ, Hallenbech JM, Kochanek P (1987) A brief episode of severe arterial hypertension induces delayed deterioration of brain function and worsens blood flow after transient multifocal cerebral ischemia. Stroke 18: 386–395

Edvinsson L, Hardebo JE, MacKenzie ET, Owman C (1978) Effects of exogenous noradrenaline on local cerebral blood flow after osmotic opening of the blood-brain barrier in the rat. J Physiol Lond 274: 149–156

– Dequeurce A, Duverger D, et al. (1983) Central serotonergic nerves project to the pial vessels of the brain. Nature 306: 55–57

Eisenberg HM, Frankowski RF, Contant CF, Marshall LF, Walker MD (1988) and the Comprehensive Central Nervous System Trauma Centers. High-dose barbiturate control of elevated intracranial pressure in patients with severe head injury. J Neurosurgery 69: 15–23

Eklöf B, Ingvar DH, Kågström E, Olin T (1971) Persistance of cerebral blood flow autoregulation following chronic bilateral cervical sympathectomy in the monkey. Acta Physiol Scand 82: 172–176

– Lassen NA, Nilsson L, Norberg K, Siesjö BK, Torlöf P (1974) Regional cerebral blood flow in the rat measured by the tissue sampling technique; a critical evaluation using four indicators C14-antipyrine, C14-ethanol, H3-water, and 133-Xenon. Acta Physiol Scand 91: 1–10

Ekström-Jodal B, Häggendal E, Linder LE, Nilsson NJ (1972) Cerebral blood flow autoregulation at high arterial pressures and different levels of carbon dioxide tension in dogs. Europ Neurol 6: 6–10

– – Johansson B, Linder LE, Nilsson NJ (1975) Acute arterial hypertension and the blood-brain barrier: An experimental study in dogs. In: Langfitt TW, McHenry LC, Reivich M, Wollmann H (eds) Cerebral circulation and metabolism. Springer, New York, pp 7–9

Ellingson I, Hauge A, Nicolaysen G, Thoresen M, Walløe L (1987) Changes in human cerebral blood flow due to step changes in PAO_2 and $PACO_2$. Acta Physiol Scand 129: 157–163

Ellis EF, Wright KF, Wei EP, Kontos HA (1981) Cyclooxygenase products of arachidonic acid metabolism in cat cerebral cortex after experimental concussive brain injury. J Neurochem 37: 892–896

Elphinstone MG, Archer DP, Pappius HM (1988) The effect of pentobarbitone anaesthesia on glucose metabolism in traumatized rat brain. Can J Anaesth: S 102–103

Enevoldsen EM, Cold GE, Jensen FT, Malmros R (1976) Dynamic changes in regional CBF, intraventricular pressure, CSF pH and lactate levels during the acute phase of head injury. J Neurosurg 44: 191–214

– Jensen FT (1978 a) Cerebrospinal fluid lactate and pH in patients with severe head injury. Clin Neurol Neurosurg 80: 213–225

– – (1978 b) Autoregulation and CO_2 responses of cerebral blood flow in patients with severe head injury. J Neurosurg 48: 689–703

– (1980) Dynamic changes in regional cerebral blood flow, cerebral ventricular pressure, cerebrospinal pH and lactate during the acute phase of severe head injury, FADL's forlag. København, Århus, Odense

– (1986) CBF in Head injury. Acta Neurochir (Wien) [Suppl] 36: 133–136

Ewing JR, Keating EG, Sheehe PR, Hodge CJ, Chipman M, Brooks CT (1981) Concordance of inhalation rCBFs with clinical evidence of cerebral ischemia. Stroke 12: 188–195

– Welch KMA, Robertson WM, Brown GG, Diaz FG, Ausman J (1983) A probe-by-probe identification of focal cerebral ischemia using the 133-Xe inhalation technique. J Cereb Blood Flow Metab 3: 586–587

Fan FC, Chen RYZ, Schuessler GB, Chien S (1979) Comparison between the 133-Xe clearance method and microsphere technique in cerebral blood flow determinations in the dog. Circ Res 44: 653–695

Farrar JK (1987) Hydrogen clearance technique. In: Wood JH (ed) Cerebral blood flow. McGraw-Hill Company, pp 275–287

Feig PU, McCurdy DK (1977) The hypertonic state. New Engl J Med 297: 1444–1454

Feruglio FS, Ruiu P, Ruiu L (1954) La protata circulatioria, il consumo di O_2 e la resistenza vascolare cerebrale dell'uomo nello stato di ibernazione artificiale con ipotermia. Minerva Med 45: 1655–1660

Feustel PJ, Ingvar MC, Severinghaus JW (1981) Cerebral oxygen availability and blood flow during middle cerebral artery occlusion: Effects of pentobarbital. Stroke 12: 858–863

Feustel PJ, Nelson LR (1987) Cortical blood flow regulation during hypoxemia in experimental head injury. J Surg Res 43: 86–93

Fieschi C, Battistini N, Beduschi A, Boselli L, Rossanda M (1974) Regional cerebral blood flow and intraventricular pressure in acute head injuries. J Neurol Neurosurg Psychiatry 37: 1378–1388

Findley G, Kohi Y, Matheson M, Teasdale G, Murray L (1983) Volumetric analysis of the contralateral dilated ventricle. J Neurol Neurosurg Psychiatry 46: 368–372

Fitch W, MacKenzie ET, Harper AM (1975) Effects of decreasing arterial blood pressure on cerebral blood flow in the baboon. Circulation Res 37: 550–557

Flamm ES, Demopoulos HB, Seligman ML, Ransohoff J (1977) Possible molecular mechanisms of barbiturate-mediated protection in regional cerebral ischaemia. Acta Neurol Scand [Suppl] 64: 150–151

Fleischer AS, Rudman DR, Fresh CB, Tindall GT (1977) Concentration of 3′, 5′ cyclic adenosine monophosphate in ventricular CSF of patients following severe head trauma. J Neurosurg 47: 517–524

Fog M (1934) Om piaarteriernes vasomotoriske reaktioner. Munksgaard, Copenhagen

Forbes HS, Nason GI, Wortman RC (1937) Cerebral circulation. Arch Neurol Psychiat 37: 334–360

Frei HJ, Wallenfang T, Pöll W, Reulen HJ, Schubert R, Brock M (1973) Regional cerebral blood flow and regional metabolism in cold induced oedema. Acta Neurochir (Wien) 29: 15–28

Freygang WH Jr, Sokoloff L (1958) Quantitative measurement of regional circulation in the central nervous system by the use of radioactive inert gas. Adv Biol Med Physics 6: 263–279

Froman C (1968) Adverse effects of low carbon dioxide tensions during mechanical over-ventilation of patients with combined head and chest injuries. Br J Anaesth 40: 383–386

Furuse M, Brock M, Hasuo M, Dietz H (1981) Relationship between brain tissue pressure gradients and cerebral blood flow distribution studied in circumscribed vasogenic cerebral oedema. Neurochirurgica 24: 10–14

Galbraith S, Cardoso E, Patterson J, Marmarou A (1984) The water content of white matter after head injury in man. In: Go KG, Baethmann AF (eds) Recent progress in the study and therapy of brain edema. Plenum Press, New York London, pp 323–330

Gennarelli TA, Obrist WD, Langfitt TW, Segawa H (1979) Vascular and metabolic reactivity to changes in PCO_2 in head injured patients. In: Popp AJ, Bourke RS, Nelson LR, Kimelberg HK (eds) Neural trauma. Raven Press, New York, pp 1–8

– (1983) Head injury in man and experimental animals: Clinical aspects. In: Hume Adams J (ed) Trauma and regeneration. Acta Neurochir [Suppl] 32: pp 1–13

– Marcincin RP, Thibault LE, Thompson CJ (1983) Effect of direction of head movement on ICP in experimental head injury. In: Ishii S, Nagai H, Brock M (eds) Intracranial pressure V. Springer, Berlin Heidelberg New York, pp 483–486

– Pastusko M, Sakamoto T, Tomei G, Duhaine A, Wiser R, Thibault L (1986) ICP after experimental diffuse head injuries. In: Miller JD, Teasdale GM, Rowan JO, Galbraith SL, Mendelow AD (eds) Intracranial pressure VI. Springer, Berlin Heidelberg New York, pp 15–19

Ginsberg MD, Graham DI, Welsh FA, Budd WW (1979) Diffuse cerebral ischemia in the cat: III. Neuropathological sequelae of severe ischemia. Ann Neurol 5: 350–358

– Smith DW, Wachtel MS, Gonzales-Carvajal M, Busto R (1986) Simultaneous determination of local cerebral glucose utilization and blood flow by carbon-14, double-label autoradiography: Method of procedure and validation studies in the rat. J Cereb Blood Flow Metab 6: 273–285

– Prado R, Dietrich WD, Busto R, Watson BD (1987) Hyperglycemia reduces the extent of cerebral infarction in rats. Stroke 18: 570–574

Gisvold SE, Safar P, Hendrick HHL, Rao G, Moossy J, Alexander H (1984) Thiopental treatment after global ischaemia in pigtailed monkeys. Anesthesiology 60: 88–96

Gjedde A, Diemer NH (1983) Autoradiographic determination of regional brain glucose content. J Cereb Blood Flow Metab 3: 303–310

Gobiet W, Grote W, Bock WJ (1975) The relation between intracranial pressure, mean arterial pressure and cerebral blood flow in patients with severe head injury. Acta Neurochir (Wien) 32: 13–24

Goldman H, Sapirstein LA (1973) Brain blood flow in the conscious and anesthetized rat. Am J Physiol 224: 122–126

Goodman SJ, Becker DP (1973) Vascular pathology of the brain stem due to experimentally increased intracranial pressure: changes noted in the micro-and macrocircualtion. J Neurosurg 39: 601–609

Gordon E, Bergvall U (1977) The effect of controlled hyperventilation on cerebral blood flow and oxygen uptake in patients with brain lesions. Acta Anaesthesiol Scand 17: 63–70

– (1979) Nonoperative treatment of acute head injuries (The Karoliska Experience). Int Anaesthesiol Clin 17: 181–199

Gotoh F, Meyer JS, Takagi Y (1965) Cerebral effects of hyperventilation in man. Arch Neurol 12: 410–423

Graham DI, Adams JH, Doyle D (1978) Ischaemic brain damage in fatal non-missile head injuries. J Neurol Sci 39: 213–234

Granholm L, Siesjö BK (1969) The effects of hypercapnia and hypocapnia upon the cerebrospinal fluid lactate and pyruvate concentrations and upon the lactate, pyruvate, ATP, ADP, phosphocreatine and creatine concentrations of the cat brain. Acta Physiol Scand 75: 257–266

Granholm L (1969) The effect of blood in the CSF on the CSF lactate, pyruvate and bicarbonate concentrations. Scand J Clin Lab Invest 23: 361–366

Gray WJ, Rosner MJ (1987) Pressure-volume index as a function of cerebral perfusion pressure. Part 2: The effects of low cerebral perfusion pressure and autoregulation. J Neurosurg 67: 377–380

Greenfield JC, Tindall GT (1968) Effect of norepinephrine, epinephrine and angiotension on blood flow in the internal carotid artery of man. J Clin Invest 47: 1672–1678

– Rembert JC, Tindall GT (1984) Transient changes in cerebral vascular resistance during the valsalva maneuver in man. Stroke 15: 76–79

Greenwood J, Luthert PJ, Pratt OE, Lantos PL (1988) Hyperosmolar opening of the blood-brain barrier in energy-depleted rat brain. Part I. Permeability studies. J Cereb Blood Flow Metab 8: 9–15

Greitz TV, Grepe AO, Kalmer MS, Lopez J (1969) Pre- and postoperative evaluation of cerebral blood flow in low-pressure hydrocephalus. J Neurosurg 31: 644–651

Gronert GA, Michenfelder JD, Sharbrough·FW, Milde JH (1981) Canine cerebral metabolic tolerance during 24 hours deep pentobarbital anesthesia. Anesthesiology 55: 110–113

Grote EH (1981) The CNS control of glucose metabolism. Acta Neurochir (Wien) [Suppl] 31: 1–160

Grote J, Zimmer K, Schubert R (1981) Effects of severe arterial hypocapnia on regional blood flow regulation, tissue PO2 and metabolism in the brain cortex of cats. Pflügers Arch 391: 195–199

Grubb RL, Raichle ME, Eichling JO, Ter-Pogossian MM (1974) The effects of changes in PaCO2 on cerebral blood volume, blood flow, and vascular mean transit time. Stroke 5: 630–638

– – Phleps ME, Ratcheson RA (1975) Effects of increased intracranial pressure on cerebral blood volume, blood flow, and oxygen utilization in monkeys. J Neurosurg 43: 385–398

Gur D, Yonas H, Jackson DL, Wolfsen SK, Roskette H, Gook WF, Maitz GS, Cook EE, Arena VC (1985) Measurement of cerebral blood flow during Xenon inhalation as measured by the microspheres method. Stroke 16: 871–874

Gyulai L, Schnall M, McLaughlin AC, Leigh JS, Chance B (1987) Silmultaneous 31P- and 1H-nuclear magnetic resonance studies of hypoxia and ischaemia in the cat brain. J Cereb Blood Flow Metab 7: 543–551

Hägerdal M, Welch FA, Keykhah MM, Perez E, Harp JR (1978) Protective effects of combinations of hypothermia and barbiturates in cerebral hypoxia in the rat. Anesthesiology 49: 165–169

Häggendal E, Johansson B (1965) Effects of arterial carbon dioxide tension and oxygen saturation on cerebral blood flow autoregulation in dogs. Acta Physiol Scand [Suppl] 258: 27–53

– – (1972) On the pathophysiology of the increased cerebrovascular permeability in acute arterial hypertension in cats. Acta Neurol Scand 48: 265–270

Haikala H, Karmalahti T, Ahtee L (1986) The nicotine-induced changes in striated dopamine metabolism of mice depend on body temperature. Brain Res 375: 313–319

Halsey J (1981) Is there clinical value in measurement of rCBF? (Editorial). rCBF Bullitin 1: 5

Hamer J, Hoyer S, Stoeckel H, Alberti E, Weinhardt F (1973) Cerebral blood flow and cerebral metabolism in acute increase of intracranial pressure. Acta Neurochir (Wien) 28: 95–110

Hankinson HL, Smith AL, Nielson SL, Hoff JT (1974) Effect of thiopental on focal cerebral ischaemia in dogs. Surg Forum 25: 445–447

Hansen NB, Nowicki PT, Miller RR, Malone T, Bickers RG, Menke JA (1986) Alterations in cerebral blood flow and oxygen consumption during prolonged hypocapnia. Pediatr Res 20: 147–150

Hardebo JE, Beley A (1984) Influence of blood pressure on bloodbrain barrier function in brain ischemia. Acta Neurol Scand 70: 356–359

Harp JR, Wollman H (1973) Cerebral metabolic effects of hyperventilation and deliberate hypotension. Br J Anaesth 45: 256–262

Harper AM, Jennett WB (1968) Simultaneous measurement of beta and gamma clearance curves of radioactive inert gases from the monkey brain. In: Bin WH, Harper AM (eds) Blood flow through organs and tissues. Livingstone Ltd, Edinburgh London, pp 214

Harrington TR, Manwaring K, Hodak J (1986) Local basal ganglia and brain stem blood flow in head injured patients using stable xenon enhanced CT scanning. In: Miller JD, Teasdale GM, Rowan JO, Galbraith SL, Mendelow AD (eds) Intracranial pressure VI. Springer, Berlin Heidelberg New York, pp 680–686

Harris RJ, Symon L, Branston NM, Bayham M (1981) Changes in extracellular calcium activity in cerebral ischaemia. J Cereb Blood Flow Metab 1: 203–209

– – (1984) Extracellular pH, potassium, and calcium activities in progressive ischaemia of rat cortex. J Cereb Blood Flow Metab 4: 178–186

– Richards PG, Symon L, Haibib AHA, Rosenstein J, (1987) pH, K+, and PO2 of the extracellular space during ischaemia of primate cerebral cortex. J Cereb Blood Flow Metab 7: 599–604

Hartmann A, Wassman H, Czernicki Z, Dettmers C, Schumacher HW, Tsuda Y (1987) Effect of stable Xenon in room air on regional cerebral blood flow and electroencephalogram in normal baboons. Stroke 18: 643–648

Hasimoto T, Pitts LH, Pogliani L, Bartkowski HM (1986) Changes in brain phosphorus metabolism in rats following fluid percussion injury. In: Miller JD, Teasdale GM, Rowan JO, Galbraith SL, Mendelow AD (eds) Intracranial pressure VI. Springer, Berlin Heidelberg New York, pp 30–34

Hass WK (1976) Prognostic value of cerebral oxidative metabolism in head trauma. In: McLaurin RL (ed) Head injuries. Grune and Stratton, New York , pp 35–37

– Rappaport ZH, Kobayashi M, Dorogi P, Hochwald GM, Ransohoff J (1976) Temporal effect of reticular formation lesions on rat cerebral blood flow (abstract). Stroke 7: 7

Hawkins RA, Hass WK, Ransohoff J (1979) Cerebral blood flow, glucose utilization, oxidative metabolism, and plasticity after mesencephalic reticular formation lesions. In: Popp AJ, Bourke RS, Nelson LR, Kimelberg HK (eds). Neural trauma. Raven Press, New York, pp 9–18

Heiss WD, Hayakawa T, Waltz AG (1976) Cortical neuronal function during ischaemia. Effects of occlusion of one middle cerebral artery on single unit activity in cats. Arch Neurol 33: 813–820

– Traupe H (1981) Comparison between hydrogen clearance and microsphere technique for rCBF measurement. Stroke 12: 161–167

– Rosner G (1983) Functional recovery of cortical neurons as related to degree and duration of ischemia. Ann Neurol 14: 294–301

Herkenham M (1981) Anesthetics and the habenulo-interpenduncular system: selective sparing of metabolic activity. Brain Res 210: 461–466

Herrschaft H, Schmidt H, Gleim F, Albus G (1975) The response of human cerebral blood flow to anesthesia with thiopentone, methohexitone, propanidide, ketamine and etomidate. In: Penzhold H, Brock M, et al (eds) Advances in neurosurgery. Springer, Berlin Heidelberg New York, pp 120–133

Hertz MM, Hemmingsen R, Bolvig TG (1977) Rapid and repetive measurements of blood flow and oxygen consumption in the rat brain using intraarterial Xenon injection. Acta Physiol Scand 101: 501–503

Hilberman M, Nioka S, Subramanian H, Egan J, Shance B (1984) Brain pH during respiratory acidoses and alkalosis, a 31P NMR study. Anesthesiology 61: A 317

Hochwald GM, Wald A, Malhan C (1976) The sink action of cerebrospinal fluid volume flow. Arch Neurol 33: 339–344

Hodes JE, Soncrant TT, Larson DM, Carlson SG, Rapoport SI (1985) Selective changes in local cerebral glucose utilization induced by phenobarbital in the rat. Anesthesiology 63: 633–639

Hoff JT, Schmith AL, Nielsen SL, Larson P (1973) Effects of barbiturate and halothane anaesthesia on focal cerebral infarction in the dog. Surg Forum 24: 449–452

– – Hankinson HL, Nielson SL (1975) Barbiturate protection from cerebral infarction in primates. Stroke 6: 28–33

Hoff JT, Pitts LH, Spetzler R, Wilson CB (1977) Barbiturates for protection from cerebral ischaemia in aneurysm surgery. Acta Neurol Scand 56: 158–159

Holm S, Vorstrup S, Lassen NA, Paulson OB (1985) Physical factors affecting calculated cerebral blood flow values in hypoperfused areas in single photon emission computerized tomography. In: Hartman A, Hoyer S (eds) Cerebral blood flow and metabolism. Springer, Berlin Heidelberg New York, pp 234–237

Holman BL, Hill TC (1987) Perfusion imaging with single-photon emission computed tomography. In: Wood JH (ed) Cerebral blood flow. Mc Graw-Hill Book Company, pp 243–256

Hoppe E, Christensen L, Christensen KN (1981) The clinical outcome of patients with severe head injuries, treated with high-dose dexamethasone, hyperventilation and barbiturates. Neurochirurgica 24: 17–20

Horton RW, Pedley TA, Meldrum BS, Chir B (1980) Regional cerebral blood flow in the rat as determined by particle distribution and by diffusible tracer. Stroke 11: 39–44

Hossmann KA (1976) Development and resolution of ischemic brain swelling. In: Pappius HM, Feindal W (eds) Dynamics of brain edema. Springer, Berlin Heidelberg New York, pp 219–228

– Sakaki S, Kimoto K (1976) Cerebral uptake of glucose and oxygen in the cat brain after prolonged ischemia. Stroke 7: 301–305

– Schuier FJ, (1980)Experimental brain infarcts in cats. Stroke 11: 583–592

Huang S-C, Carson RE, Phleps ME (1982) Measurement of local blood flow and distribution volume with short-lived isotopes: A general input technique. J Cereb Blood Flow Metab 2: 99–108

Høedt-Rasmussen K, Sveinsdottir E, Lassen NA (1966) Regional cerebral blood flow in man determined by intraarterial injection of radioactive inert gas. Circ Res 18: 237–247

– (1967) Regional blood flow. The intra-arterial injection method. Acta Neurol Scand 43 [Suppl] 27: 1–81

Inao S, Marmarou A, Clarke GD, Andersen BJ, Fatouros PP, Young HF (1988) Production and clearance of lactate from brain tissue, cerebrospinal fluid, and serum following experimental brain injury. J Neurosurg 69: 736–744

Ingvar DH, Häggendal E, Nilsson E, Sourander NJ, Wickbom J, Lassen NA (1964) Cerebral circulation and metabolism in a comatose patient, studied with a new method. Arch Neurol 11: 13–19

– Sourander P (1970) Destruction of the reticular core of the brain stem. Arch Neurol 23: 1–8

– Brun A (1972) Das komplette apallische syndrom. Arch Psychiat Nervenkr 215: 219–239

– Lassen NA (1973) Cerebral complications following measurements of regional cerebral blood flow (rCBF) with the intraarterial 133-Xenon injection method. Stroke 4: 658–665

– – (eds) Brain work. Copenhagen, Munksgaard 1975

– – (1982) Atraumatic two-dimensional rCBF measurements using stationary detectors and inhalation or intravenous administration of 133-xenon. J Cereb Blood Flow Metab 2: 271–274

Ishige N, Pitts LH, Berry I, Carlson SG, Nishimura MC, Moseley ME, Weinstein PR (1987) The effect of hypoxia on traumatic head injury in rats: Alterations in neurologic function, brain edema, and cerebral blood flow. J Cereb Blood Flow Metab 7: 759–767

– – – Nishimura MC, James TL (1988) The effects of hypovolemic hypotension on high-energy phosphate metabolism of traumatized brain in rats. J Neurosurg 68: 129–136

Ito U, Spatz M, Walker JT, Klatzo I (1975) Experimental cerebral ischaemia in mongolian gerbils. I. Light microscopic observations. Acta Neuropathol (Berl) 32: 209–223

– Ohno K, Yamaguchi T, Takei H, Tomita H, Inaba Y (1980) Effect of hypertension on blood-brain barrier change after restoration of blood flow in post-ischemic gerbil brains. Stroke 11: 606–611

Jafar JJ, Johns LM, Mullan SF (1986) The effect of mannitol on cerebral blood flow. J Neurosurg 64: 754–759

Jaggi JL, Obrist WD, Gennarelli TA, Langfitt TW (1990) Relationship of early cerebral blood flow and metabolism to outcome in acute head injury. J Neurosurg 72: 176–182

James HE, Bruce DA, Welch F (1978) Cytotoxic edema produced by 6-Aminonicotinamide and its response to therapy. Neurosurgery 3: 196–200

– (1980) Methodology for the control of intracranial pressure with hypertonic mannitol. Acta Neurochir (Wien) 51: 161–172

Jennett B, Teasdale C (1981) Dynamic pathology. In: Jennett B, Teasdale C (eds) Management of head injury. Davis Company, Philadelphia, pp 45–75

Johansson BB, Li C, Olsson Y, Klatzo I, (1970) The effect of acute arterial hypertension on the blood-brain barrier to protein tracers. Acta Neuropathol (Berl) 16: 117–124

– (1974) Blood brain barrier disfunction in acute arterial hypertension after papaverine-induced vasodilatation. Acta Neurol Scand 50: 573–580

– Linder LE (1978) Reversibility of the blood-brain barrier disfunction induced by acute hypertension. Acta Neurol Scand 57: 345–348

– Auer LM (1983) Neurogenic modification of the vulnerability of the blood-brain barrier during acute hypertension in conscious rats. Acta Physiol Scand 117: 507–511

Johnston IH, Johnston JA, Jennett B (1970) Intracranial pressure changes following head injury. Lancet ii: 433–436

– Rowan JO, Harper AM, Jennett WB (1972) Raised intracranial pressure and cerebral blood flow. 1: cisterna magna infusion in primates. J Neurol Neurosurg Psychiatry 35: 285–296

– – – (1973) Raised intracranial pressure and cerebral blood flow. 2: Supratentorial and infratentorial mass lesions in primates. J Neurol Neurosurg Psychiatry 36: 161–170

– Harper AM (1973) The effect of mannitol on cerebral blood flow. An experimental study. J Neurosurg 38: 461–471

– Rowan JO (1974) Raised intracranial pressure and cerebral blood flow 4: Intracranial pressure gradients and regional cerebral blood flow. J Neurol Neurosurg Psychiatry 37: 585–592

Jones TH, Morawetz RB, Crowell RM, Marcoux FW, FitzGibbon SJ, Degirolami U, Ojemann RG (1981) Thresholds of focal cerebral ischemia in awake monkeys. J Neurosurg 54: 773–782

Junck L, Dhawan V, Thaler HT, Rottenberg DA (1985) Effects of Xenon and Krypton on regional cerebral blood flow in the rat. J Cereb Blood Flow Metab 5: 126–132

Kanno I, Lassen NA (1979) Two methods for calculating regional cerebral blood flow from emission computed tomography of inert gas concentrations. J Comput Assist Tomogr 3: 71–76

Kasoff SS, Zingesser LH, Shulman K (1972) Compartmental abnormalities of regional cerebral blood flow in children with head trauma. J Neurosurg 36: 463–470

Kassell NF, Hitchon PW, Gerk MK, Sokoll MD, Hill TR (1980) Alterations in cerebral blood flow, oxygen metabolism, and electrical activity produced by high dose sodium thiopental. Neurosurgery 7: 598–603

– – – – – (1981) Influence of changes in arterial pCO_2 on cerebral blood flow and metabolism during high-dose barbiturate therapy in dogs. J Neurosurg 54: 615–619

– Baumann KW, Hitchon PW, Gerk MK, Hill TR, Sokoll MD (1982) The effects of high dose mannitol on cerebral blood flow in dogs with normal intracranial pressure. Stroke 13: 59–61

Kaste M, Hernesniemi J, Somer H, Hillbom M, Konttinen A (1981) Creatine kinase isoenzymes in acute brain injury. J Neurosurg 55: 511–515

Katz JM, Abou-Madi M, Abou-Madi N, Trop D (1986) Do mannitol-induced haemodynamic responses influence its effect on intracranial pressure? A study in the dog with and without induced intracranial hypertension. Can Anaesth Soc J 33: S 81–S 82

Kennealy JA, McLennan JE, Loudon G, McLaurin RL (1980) Hyperventilation-induced cerebral hypoxia. Am Rev Resp Dis 122: 407– 412

Kennedy C, Sokoloff L (1957) An Adaptation of the nitrous oxide method to the study of the cerebral circulation in children; normal values for cerebral blood flow and cerebral metabolic rate in children. J Clin Invest 36: 1130–1137

Kety SS, Schmidt CF (1945) The determination of cerebral blood flow in man by the use of nitrous oxide in low concentrations. Am J Physiol 143: 53–66

– – (1948) The nitrous oxide method for the quantitative determination of cerebral blood flow in man: Theory, procedure and normal values. J Clin Invest 27: 476–483

Keykhah MM, Hägerdal M, Welsh FA, Barrer MA, Sisco F, Harp JR (1980) Effect of high vs. low arterial blood oxygen content on cerebral energy metabolite levels during hypoxia with normothermia and hypothermia in the rat. Anesthesiology 52: 492–495

King BD, Sokoloff L, Wechsler L (1952) The effect of l-epinephrine and l-nor-epinephrine upon cerebral circulation and metabolism in man. J Clin Invest 31: 273–279

King LJ, Lowry OH, Passonneau JV, Venson V (1967) Effects of convulsants on energy reserves in the cerebral cortex. J Neurochem 14: 599–611

King LR, McLaurin RL, Knowles HC (1974) Acid-base balance and arterial and CSF lactate levels following human head injury. J Neurosurg 40: 617–625

Kitahata LM, Galicich JM, Sato I (1971) The effect of passive hyperventilation on intracranial pressure in the dog. J Neurosurg 34: 185–193

Klauber MR, Toutant SM, Marshall LF (1984) A model for predicting delayed intracranial hypertension following head injury. J Neurosurg 61: 695–699

Kleinerman J, Sancetta SM, Hackel DB (1958) Effects of high spinal anaesthesia on cerebral circulation and metabolism in man. J Clin Invest 37: 285–293

Kobayashi S, Nakazawa S, Yano M, Yamamoto Y, Otsuka T (1983) The value of intracranial pressure (ICP) measurement in acute severe head injury showing diffuse cerebral swelling. In: Ishii S, Nagai H, Brock M (eds) Intracranial pressure V. Springer, Berlin Heidelberg New York, pp 527–531

Koch KA, Jackson DL, Schmiedl M, Thompson WL, Rosenblatt JI (1984) Effect of thiopental therapy on cerebral blood flow after total cerebral ischemia. Crit Care Med 12: 90–95

Kofke WA, Nemoto EM, Hossmann KA, Taylor F, Kessler PD, Stezoski SW (1979) Brain blood flow and metabolism after global ischemia and post-insult thiopental therapy in monkeys. Stroke 10: 554–560

Kontos HA, Wei EP, Navari RM, Levasseur JE, Rosenblum WI, Patterson JL (1978) Responses of cerebral arteries and arterioles to acute hypotension and hypertension. Am J Physiol 234: H 371–383

– – Dietrich WD, Navari RM, Povlishock JT, Ghatak NR, Ellis EF, Patterson JL (1981) Mechanism of cerebral arteriolar abnormalities after acute hypertension. Am J Physiol 240: H 511–527

– – (1986) Superoxide production in experimental brain injury. J Neurosurg 64: 803–807

Kopf GS, Mirvis DM, Myers RE (1975) Central nervous system tolerance to cardiac arrest during profound hypothermia. J Surg Res 18: 29–34

Kosteljanetz M (1986) Acute head injury: Pressure-volume relations and cerebrospinal fluid dynamics. Neurosurgery 18: 17–24

Kraig RP, Chesler M (1987) Glial acid-base behaviour in brain ischaemia (abstract). J Cereb Blood Flow Metab 7 [Suppl] 1: S 126

– Pepito CK, Plum F, Pulsinelli WA (1987) Hydrogen ions kill brain at concentrations reached in ischemia. J Cereb Blood Flow Metab 7: 379–386

Kramer RS, Sanders AP, Lesage AM et al. (1968) The effect of profound hypothermia on preservation of cerebral ATP content during circulatory arrest. J Thorac Cardiovasc Surg 56: 699–709

Krantis A (1983) Hypothermia-induced reduction in the permeation of the radiolabelled tracer substances across the blood-brain barrier. Acta Neuropathol (Berl) 60: 61–69

Krieglstein J, Sperling G, Twietmeyer G (1981) Effects of thiopental on regulatory mechanisms of brain energy metabolism. Acta Pharmacol 318: 56–61

Kuhl DE, Alavi A, Hoffman EJ, Phleps ME, Zimmerman RA, Obrist WD, Bruce DA, Greenberg JH, Uzzell B (1980) Local cerebral blood volume in head-injured patients. J Neurosurg 52: 309–320

Kurze T, Tranquada RE, Benedict K (1966) Spinal fluid lactic acid levels in acute cerebral injury. In: Caveness WF, Walker AE (eds) Head injury. JB Lippincott, Philadelphia, pp 254–259

Kuschinsky W, Suda S, Bunger R, Sokoloff L (1982) The influence of L-norepinephrine on the local coupling between brain metabolism and blood flow. In: Heistad DD, Marcus ML (eds) Cerebral blood flow: effects of nerves and neurotransmitters. Elsevier, North-Holland, New York, pp 169–176

Lafferty JJ, Keykhah MM, Shapiro HM, VanHorn K, Behar MG (1978) Cerebral hypometabolism obtained with deep pentobarbital anesthesia and hypothermia (30 °C). Anesthesiology 49: 159–164

LaMorgese J, Fein JM, Shulman K (1975) Polarographic and microsphere analysis of ultraregional cerebral blood flow rates in the cat. In: Harper AM, Jennett WB, Miller JD, Rowan JO (eds) Blood flow and metabolism in the brain. Churchill Livingstone, Edinburgh, pp 7.3–7.8

Landau WM, Freygang WH Jr, Roland LP, Sokoloff L, Kety SS (1955) The local circulation of the living brain; values in the unanesthetized and anesthetized cat. Trans Am Neurol Assoc 80: 125–129

Langfitt TW, Obrist WD, Gennarelli TA, O'Connor MJ, Weeme AT (1977) Correlation of cerebral blood flow with outcome in head injured patients. Ann Surg 186: 411–414

– Gennarelli TA (1982) Can the outcome from head injury be improved? J Neurosurg 56: 19–25

Lantos J, Temes G, Torok B (1986) Changes during ischaemia in extracellular potassium ion concentration of the brain under nitrous oxide or hexobarbital-sodium anesthesia and moderate hypothermia. Acta Physiol Scand 67: 141–153

Larsen B, Skinhøj E, Soh K, Endo H, Lassen NA (1977) The pattern of cortical activity provoked by listening and speech revealed by rCBF measurements. Acta Neurol Scand 56 [Suppl] 64: 268–269

Lassen NA, Munck O (1955) The cerebral blood flow in man determined by the use of radioactive Krypton. Acta Physiol Scand 33: 30–49

– (1959) Cerebral blood flow and oxygen consumption in man. Physiol Rev 39: 183–238

– Feinberg I, Lane MH (1960) Bilateral studies of cerebral oxygen uptake in young and aged normal subjects and in patients with organic dementia. J Clin Invest 39: 491–500

– Lane MH (1961) Validity of internal jugular blood for study of cerebral blood flow and metabolism. J Appl Physiol 16: 313–320

– Ingvar DH (1963) Regional cerebral blood flow measurement in man. Arch Neurol 9: 615–622

– (1974) Control of cerebral circulation in health and disease. Circ Res 34: 749–760

– Christensen MS (1976) Physiology of cerebral blood flow. Br J Anesth 48: 719–735

– Henriksen L, Paulson O (1981) Regional cerebral blood flow in stroke by 133-xenon inhalation and emission tomography. Stroke 12: 284–288

– (1986) Cerebral and spinal cord blood flow. In: Cottrell JE, Turndorff H (eds) Anaesthesia and neurosurgery. Mosby Company, St. Louis, Toronto, pp 1–22

– Astrup J (1987) Ischaemic penumbra. In: Wood JH (ed) Cerebral blood flow, physiologic and clinical aspects. McGraw-Hill Book Company, pp 458–466

Laurent JP, Lawner P, Simeone FA, Fink E (1982) Pentobarbital changes compartmental contribution to cerebral blood flow. J Neurosurg 56: 504–510

Lear JL, Jones SC, Greenberg JH, Fedora TJ, Reivich M (1981) Use of 123-I and 14-C in a double radionuclide autoradiographic technique for simultaneous measurement of LCBF and LCMRgl: Theory and method. Stroke 12: 589–597

– Ackermann R, Kamayama M, Carson R, Phleps M (1984) Multiple-radionuclide autoradiography in evaluation of cerebral function. J Cereb Blood Flow Metab 4: 264–269

Leech PJ, Miller JD, Fitch W, Barker J (1974) Cerebral blood flow, internal carotid artery pressure, and EEG as a guide to the safety of carotid ligation. J Neurol Neurosurg Psychiatry 37: 854–862

Levin HS, Williams D, Crofford MJ, High WM, Eisenberg HM, Amparo EG, Guinto FC, Kalisky Z, Handel SF, Goldman AM (1988) Relationship of depth of brain lesions to consciousness and outcome after closed head injury. J Neurosurg 69: 861–866

Levy DE, Duffy TE (1977) Cerebral energy metabolism during transient ischaemia and recovery in the gerbil. J Neurochem 28: 63–70

– Brierley JB (1979) Delayed pentobarbital administration limits ischemic brain damage in gerbils. Ann Neurol 5: 59–64

Lewelt W, Jenkins LW, Miller JD (1980) Autoregulation of cerebral blood flow after experimental fluid percussion injury of the brain. J Neurosurg 53: 500–511

– – (1982) Effect of experimental fluid-percussion injury of the brain on cerebrovascular reactivity to hypoxia and to hypercapnia. J Neurosurg 56: 332–338

Litt L, Gonzáles-Mendez R, Severinghaus JW, Hamilton WK, Shuleshko J, Murphy-Boesch J, James TL (1985) Cerebral intra-

cellular changes during supercarbia: An in vivo 31P nuclear magnetic resonance study in rats. J Cereb Blood Flow Metab 5: 537–544

Little JR (1978) Modification of acute focal ischemia by treatment with mannitol. Stroke 9: 4–9

Lundberg N, Kjällquist A, Bien C (1959) Reduction of increased intracranial pressure by hyperventilation. Acta Psychiat Neurol Scand 34 (suppl 139)

– (1960) Continuous recording and control of ventricular fluid pressure in neurosurgical practice. Acta Psychiat Scand 36 (suppl 149)

Maas AIR (1977) Cerebrospinal fluid enzymes in acute brain injury. Dynamics of changes in CSF enzyme activity after acute experimental brain injury. J Neurol Neurosurg Psychiatry 40: 655–665

MacKenzie ET, McCulloch J, O'Keane M, Pickard JD, Harper AM (1976 a) Cerebral circulation and norepinephrine: relevance of the blood-brain barrier. Am J Physiol 231: 483–488

– Strandgaard S, Graham DI, Jones JV, Harper AM, Farrar JK (1976 b) Effects of acutely induced hypertension in cats on pial arteriolar caliber, local cerebral blood flow, and the blood-brain barrier. Circ Res 39: 33–41

– McGeorge AD, Graham DT et al. (1979) Effects of increasing arterial pressure on cerebral blood flow in the baboon: Influence of the sympathetic nervous system. Pflügers Arch 378: 189–195

MacMillan V, Siesjö BK (1973) The effect of phenobarbitone anesthesia upon some organic phosphates, glycolytic metabolites and citric acid cycle-associated intermediates of the rat brain. J Neurochem 20: 1669–1681

Makowski EL, Meschia G, Droegenmueller W, Battaglia FC (1968) Measurement of umbilical arterial blood flow to the sheep placenta and fetus in utero. Circ Res 23: 623–631

Mallett BL, Veall N (1963) Investigation of cerebral blood flow in hypertension, using radioactive-xenon inhalation and extracranial recording. Lancet i: 1081–1082

Maier-Hauff K, Baethmann AJ, Lange M, Schürer L, Unterberg A (1984) The kallikrein-kinin system as mediator in vasogenic brain edema. J Neurosurg 61: 97–106

Malmlund HO, Lying-Tunell U, Böhmer G (1972) The effect of ventricular-atrial shunting on cerebral oxygen consumption in patients with dementia. Europ Neurol 6: 340–345

Mangold R, Sokoloff L, Conner E, Kleinerman J, Therman PG, Kety SS (1955) The effects of sleep and lack of sleep on the cerebral circulation and metabolism of normal young men. J Clin Invest 34: 1092–1100

Marbach EP, Weil MH (1967) Rapid enzymatic measurement of blood lactate and pyruvate. Use and significance of metaphosphorate as a common precipitant. Clin Chem 13: 314–325

Marcoux FW, Morawetz RB, Crowell RM, DeGirolami U, Halsey JH (1982) Differential regional vulnerability in transient focal cerebral ischemia. Stroke 13: 339–346

Marcus ML, Bischof CJ, Heistad DD (1981) Comparison of microsphere and 133 Xenon clearance method in measuring skeletal muscle and cerebral blood flow. Circ Res 48: 748–761

Marmarou A, Maset AL, Ward JD, Moulton RJ, Lutz HA, Clifton GL, Becker DP (1986) Dynamics of intracranial pressure rise in severely injured patients. In: Miller JD, Teasdale GM, Rowan JO, Galbraith SL, Mendelow AD (eds) Intracranial pressure VI. Springer, Berlin Heidelberg New York, pp 9–14

– – – Choi S, Brooks D, Lutz HA, Moulton RJ, Muizelaar JP, DeSalles A, Young HF (1987) Contribution of CSF and vascular factors to elevation of ICP in severely head-injured patients. J Neurosurg 66: 883–890

Marshall SB, Owens JC, Swan H (1956) Temporary circulatory occlusion to the brain of the hypothermic dog. Arch Surg 72: 98–106

Marshall LF, Durity F, Lounsbury R, Graham DI, Welch F, Langfitt TW (1975 a) Experimental cerebral oligemia and ischemia produced by intracranial hypertension Part 1: Pathophysiology, electroencephalography, cerebral blood flow, blood-brain barrier, and neurological function. J Neurosurg 43: 308–317

– Graham DI, Path MRC, Durity F, Lounsbury R, Welch F, Langfitt TW (1975 b) Experimental cerebral oligemia and ischemia produced by intracranial hypertension Part 2: Brain morphology. J Neurosurg 43: 318–322

– Welch F, Durity F, Lounsbury R, Graham DI, Path MRC, Langfitt TW (1975 c) Experimental cerebral oligemia and ischemia produced by intracranial hypertension Part 3: Brain energy metabolism. J Neurosurg 43: 323–328

– Smith RW, Rauscher LA, Shapiro HM (1978) Mannitol dose requirements in brain-injured patients. J Neurosurg 48: 169–172

– – Shapiro HM (1979 a) The outcome with aggressive treatment in severe head injuries Part I: The significance of intracranial pressure monitoring. J Neurosurg 50: 20–25

– – – (1979 b) The outcome with aggressive treatment in severe head injuries Part II: Acute and chronic barbiturate administration in the management of head injury. J Neurosurg 50: 26–30

Martins AN, Doyle TF, Newby N (1976) PCO$_2$ and rate of formation of cerebrospinal fluid in monkey. Am J Physiol 231: 127–131

Maset AL, Marmarou A, Ward JD, Choi S, Lutz HA, Brooks D, Moulton RJ, DeSalles A, Muizelaar JP, Turner H, Young HF (1987) Pressure-volume index in head injury. J Neurosurg 67: 832–840

Mayhan WG, Heistad DD (1986) Role of veins and cerebral venous pressure in disruption of the blood-brain barrier. Circ Res 59: 216–220

– Faraci FM, Heistad DD (1986) Disruption of the blood-brain barrier in cerebrum and brain stem during acute hypertension. Am J Physiol 251: H 1171–1175

McCulloch J, Savaki HE, Jehle J, Sokoloff L (1982) Local cerebral glucose utilization in hypothermia and hyperthermic rats. J Neurochem 39: 225–258

McHenry LC, Slocum HC, Bivens HE, Mayes HA, Hayes GJ (1965) Hyperventilation in awake and anesthetized man. Arch Neurol 12: 270–277

McIlwain H, Poll JD (1986) Adenosine in cerebral homeostatic role: appraisal through actions of homocysteine, colchicine and dipyridamole. J Neurobiol 17: 39–49

McKay RD, Sundt TM, Michenfelder JD, Gronert GA, Messick JM, Sharbrough FW, Piepgras DG (1976) Internal carotid artery stump pressure and cerebral blood flow during carotid endarterectomy: Modification by halothane, enflurane, and innovar. Anaesthesiology 45: 390–399

McKissock W, Paine KWE, Walsh LS (1960) The value of hypothermia in the surgical treatment of ruptured intracranial aneurysms. J Neurosurg 17: 700–707

McKrell TN, Stone HH, Wechsler RL (1955) Effect of drug induced hypotension on the cerebral circulation in man. Surg Forum 5: 730–736

McQueen JD, Jeanes LD (1964) Dehydration and rehydration of the brain with hypertonic urea and mannitol. J Neurosurg 21: 118–128

Meldrum BS, Nilsson B (1976) Cerebral blood flow and metabolic rate early and late in prolonged epileptic seizures induced in the rat by bicuculline. Brain 99: 523–543

Mendell PL, Hollenberg NK (1971) Cardiac output distribution in the rat. Comparison of rubidium and microsphere methods. Am J Physiol 221: 1617–1620

Mendelow AD, Teasdale GM, Teasdale E, Matheson M, Russell T (1983) Cerebral blood volume and intracranial pressure in head injured patients. In: Ishii S, Nagai H, Brock M (eds) Intracranial pressure V. Springer, Berlin Heidelberg New York, pp 495–500

– – Russell T, Flood J, Patterson J, Murray GD (1985) Effect of mannitol on cerebral blood flow and cerebral perfusion pressure in human head injury. J Neurosurg 63: 43–48

Messeter K, Pontén U, Siesjö BK (1972) The influence of deep barbiturate anaesthesia upon the regulation of extra- and intracellular pH in the rat brain during hypercapnia. Acta Physiol Scand 85: 174–182

– Nordström C-H, Sundbärg G, Algotsson L, Ryding E (1986) Cerebral hemodynamics in patients with severe head trauma. J Neurosurg 64: 231–237

Messick JM, Casement B, Sharbrough FW, Milde LN, Michenfelder JD, Sundt TM (1987) Correlation of regional cerebral blood flow (rCBF) with EEG changes during isoflurane anesthesia for carotid endarterectomy: Critical rCBF. Anesthesiology 66: 344–349

Metzel E, Zimmermann WE (1971) Changes of oxygen pressure, acid-base balance, metabolites and electrolytes in cerebrospinal fluid and blood after cerebral injury. Acta Neurochir (Wien) 25: 177–188

Meyer FB, Anderson RE, Sundt TM, Yakch TL (1987) Treatment of experimental focal cerebral ischaemia with mannitol. J Neurosurg 66: 109–115

Meyer JS, Hayman LH, Yamamoto M, Sakal F, Nakajima S (1980) Local cerebral blood flow measured by CT after stable Xenon inhalation. Am J Radiol 135: 239–251

Michenfelder JD, Theye RA (1968) Hypothermia: Effect on canine brain and whole-body metabolism. Anesthesiology 29: 1107–1112

– Messick CM, Theye RA (1968) Simultaneous cerebral blood flow measured by direct and indirect methods. J Surg Res 8: 475–481

– VanDyke RA, Theye RA (1970) The effects of anesthetic agents and techniques on canine cerebral ATP and lactate levels. Anesthesiology 33: 315–321

– Theye RA (1970) The effects of anesthesia and hypothermia on canine cerebral ATP and lactate during anoxia produced by decapitation. Anesthesiology 33: 430–439

– Theye RA (1973) Cerebral protection by thiopental during hypoxia. Anesthesiology 39: 510–517

– (1974) The interdependency of cerebral functional and metabolic effects following massive doses of thiopental in the dog. Anesthesiology 41: 231–236

– Milde JH, Sundt JM jr (1976) Cerebral protection by barbiturate anesthesia. Use after middle cerebral artery occlusion in Jave monkeys. Arch Neurol 33: 345–350

– – (1977) Failure of prolonged hypocapnia, hypothermia or hypertension to favourably alter acute stroke in primates. Stroke 8: 87–91

Mies G, Kliber O, Drewes LR, Hossmann K-A (1984) Cerebral blood flow and regional potassium distribution during focal ischemia of gerbil brain. Ann Neurol 16: 232–237

Miller JD, Ledingham IM, Jennett WB (1970) Effects of hyperbaric oxygen on intracranial pressure and cerebral blood flow in experimental cerebral oedema. J Neurol Neurosurg Psychiatry 33: 745–755

– Leech P (1975) Effects of mannitol and steroid therapy on intracranial volume-pressure relationships in patients. J Neurosurg 42: 274–281

– Becker DP, Ward JD, Sullivan HG, Adams WE, Rosner MJ (1977) Significance of intracranial hypertension in severe head injury. J Neurosurg 47: 503–516

– (1979) Barbiturates and raised intracranial pressure. Ann Neurol 6: 189–193

– Butterworth JF, Gudeman SK, Faulkner JE, Choi SC, Selhorst JB, Harbison JW, Lutz HA, Young HF, Becker DP (1981) Further experience in the management of severe head injury. J Neurosurg 54: 289–299

Mitchell DS, Adams JH (1973) Primary focal impact damage to the brainstem in blunt head injuries. Does it excist? Lancet ii: 215–218

Morawetz RB, Marcoux FW, Crowell RM, DeGirolami U, Halsey JH (1979) Identical thresholds of cerebral ischemia in white and grey matter. Acta Neurol Scand 60 [Suppl] 72: 282–283

Morgan P, Ward B (1970) Hyperventilation and changes in the electroencephalogram and electroretinogram. Neurology 20: 1009–1014

Morii S, Ngai AC, Ko KR, Winn HR (1986 a) A venous outflow method for continuously monitoring cerebral blood flow in the rat. Am J Physiol 250: H 304–H 312

– – Winn HR (1986 b) Reactivity of rat pial arterioles and venules to adenosine and carbon dioxide: With detailed description of the closed cranial window technique in rats. J Cereb Blood Flow Metab 6: 34–41

Moss E, Gibson SJ, McDowall DG, Gibson RM (1983) Intensive management of severe head injuries. Anaesthesia 38: 214–225

Muizelaar JP, Wei EP, Kontos HA, Becker DP (1983) Mannitol causes compensatory cerebral vasoconstriction and vasodilatation in response to blood viscosity changes. J Neurosurg 59: 822–828

– Lutz HA, Becker DP (1984) Effect of mannitol on ICP and CBF and correlation with pressure autoregulation in severely head-injured patients. J Neurosurg 61: 700–706

– van Der Poel HG, Li Z, Kontos HA, Levasseur JE (1988) Pial arteriolar vessel diameter and CO_2 reactivity during prolonged hyperventilation in the rabbit. J Neurosurg 69: 923–927

– Marmarou A, DeSalles AAF, Ward JD, Zimmerman RS, Li Z, Choi SC, Young HF (1989 a) Cerebral blood flow and metabolism in severely head-injured children. Part I: Relationship with GCS score, outcome, ICP, and PVI. J Neurosurg 71: 63–71

– Ward JD, Marmarou A, Newlon PG, Wachi A (1989 b) Cerebral blood flow and metabolism in severely head-injured children. Part 2: Autoregulation. J Neurosurg 71: 72–76

Murphy A, Teasdale E, Matheson M, Galbraith S, Teasdale G (1983) Relationship between CT indices of brain swelling and intracranial pressure after head injury. In: Ishii S, Nagai S, Brock M (eds) Intracranial pressure V. Springer, Berlin Heidelberg New York, pp 562–566

Murphy VA, Johanson CE (1985) Adrenergic-induced enhancement of brain barrier system permeability to small nonelectrolytes: choroid plexus versus cerebral capillaries. J Cereb Blood Flow Metab 5: 401–412

Murray IPC, Hoschl R, Choy D (1978) The jugular venous reflux. Clin Nucl Med 3: 56–57

Myers RE, Yamaguchi S (1977) Nervous system effects of cardiac arrest in monkeys. Arch Neurol 34: 65–74

Nagai H, Yamamoto L, Diksic M, Worsley KJ, Takara E (1988) Tripple-tracer autoradiography demontrates effects of hyperglycemia on cerebral blood flow , pH, and glucose utilization in cerebral ischemia of rats. Stroke 19: 764–772

Narayan RK, Kishore PRS, Becker DP, Ward JD, Enas GG, Greenberg RP, Silva AD, Lipper MH, Choi SC, Mayhaml CG, Lutz HA, Young HF (1982) Intracranial pressure: to monitor or not to monitor? A review of our experience with severe head injury. J Neurosurg 56: 650–659

Naritomi H, Sasaki M, Kanashiro M, Kitani M, Sawada T (1988) Flow thresholds for cerebral energy disturbance and Na + pump failure as studied by in vivo 31P and 23 Na nuclear magnetic resonance spectroscopy. J Cereb Blood Flow Metab 8: 16–23

Nath F, Galbraith S (1986) The effect of mannitol on cerebral white matter water content. J Neurosurg 65: 41–43

Nedergaard M, Gjedde A, Diemer NH (1986) Focal cerebral ischaemia of the rat brain: Autoradiographic determination of cerebral glucose utilization, glucose content and blood flow. J Cereb Blood Flow Metab 6: 414–424

– Diemer NH (1987) Focal ischemia of the rat brain, with special reference to the influence of plasma glucose concentration. Acta Neuropathol (Berl) 73: 131–137

– (1988) Mechanisms of brain damage in focal cerebral ischaemia. Acta Neurol Scand 77: 1–24

– Jakobsen J, Diemer NH (1988) Autoradiographic determination of cerebral glucose content, blood flow, and glucose utilization in focal ischemia of the rat brain: Influence of the plasma glucose concentration. J Cereb Blood Flow Metab 8: 100–108

Neill WA, Hattenhauer M (1975) Impairment of myocardial O_2 supply due to hyperventilation. Circulation 52: 854–858

Nilsson B, Nordström C-H (1977 a) Experimental head injury in the rat. Part 3: Cerebral blood flow and oxygen consumption after concussive impact acceleration. J Neurosurg 47: 262–273

– – (1977 b) Rate of cerebral energy consumption in concussive head injury in the rat. J Neurosurg 47: 274–281

– Pontén U, Voigt G (1977) Experimental head injury in the rat. Part 1: Mechanics, pathophysiology, and morphology in an impact acceleration trauma model. J Neurosurg 47: 241–251

– – (1977) Experimental head injury in the rat. Part 2: Regional brain energy metabolism in concussive trauma. J Neurosurg 47: 252–261

– Siesjö BK (1983) A venous outflow method for measurement of rapid changes of cerebral blood flow and oxygen consumption in the rat. Stroke 14: 797–802

Nilsson L (1971) The influence of barbiturate anaesthesia upon the energy state and upon acid-base parameters of the brain in arterial hypotension and in asphyxia. Acta Neurol Scand 47: 233–253

– Siesjö BK (1974) Influence of anaesthesia on the balance between production and utilization of energy in the brain. J Neurochem 23: 29–36

– – (1975) The effect of phenobarbitone anaesthesia on blood flow and oxygen consumption in the rat brain. Acta Anaesthesiol Scand [Suppl] 57: 18–24

Nordby HK, Urdal P (1982) The diagnostic value of measuring creatine kinase BB activity in cerebrospinal fluid following acute head injury. Acta Neurochir (Wien) 65: 93–101

Nordström C-H, Rehncrona S (1977) Postischaemic cerebral blood flow and oxygen utilization rate in rats anaesthetized with nitrous oxide or phenobarbital. Acta Physiol Scand 101: 230–240

– Siesjö BK (1978) Influence of phenobarbital on changes in the metabolites of the energy reserve of the cerebral cortex following complete ischaemia. Acta Physiol Scand 104: 271–280

– Rehncrona S, Siesjö BK (1978) Effect of phenobarbitone in cerebral ischemia. Part II: Restitution of cerebral energy state, as well as of glycolytic metabolites, citric acid cycle intermediates and associated aminoacids after pronounced imcomplete ischemia. Stroke 9: 335–343

– – (1978) Reduction of cerebral blood flow and oxygen consumption with a combination of barbiturate anaesthesia and induced hypothermia in the rat. Acta Anaesthesiol Scand 22: 7–12

– Messeter K, Sundbärg G, Schalén W, Werner M, Ryding E (1988) Cerebral blood flow, vasoreactivity and oxygen consumption during barbiturate therapy in severe traumatic brain lesions. J Neurosurg 68: 424–431

Norwood WI, Norwood CR (1982) Influence of hypothermia on intracellular pH during anoxia. Am J Physiol 243: 62–65

Nuetze JM, Wyler F, Rudolph AM (1968) Use of radioactive microspheres to assess distribution of cardial output in rabbits. Am J Physiol 215: 486–495

Obrist WD, Thompson HK, King CH, Wang HS (1967) Determination of regional cerebral blood flow by inhalation of 133-Xenon. Circ Res 20: 124–135

– Gennarelli TA, Segewa H, Dolinskas CA, Langfitt TW (1979) Relation of cerebral blood flow to neurological status and outcome in head-injured patients. J Neurosurg 51: 292–300

– Langfitt TW, Jaggi JL, Cruz J, Gennarelli TA (1984) Cerebral blood flow and metabolism in comatose patients with acute head injury. Relationship to intracranial hypertension. J Neurosurg 61: 241–253

– Jaggi JL, Harel D, Smith DS (1985) Effect of stable Xenon inhalation of human CBF. J Cereb Blood Flow Metab 5: 557–558

– Wilkinson WE (1985) Stability and sensitivity of CBF indices in the noninvasive 133-Xe method. In: Hartmann A, Hoyer S (eds) Cerebral blood flow and metabolism measurement. Springer, Berlin Heidelberg New York, pp 30–36

Okada Y, Shima T, Yamamoto M, Uozumi J (1983) Regional cerebral blood flow, sensory evoked potentials, and intracranial pressure in dogs with MCA occlusion by embolization or trapping. J Neurosurg 58: 500–507

Okuda C, Saito A, Miyazaki M, Kuriyama K (1986) Alteration of the turnover of dopamine and 5-hydroxytryptamine in rat brain associated with hypothermia. Pharmacol Biochem Behav 24: 79–83

Olesen J (1971) Contralateral focal increase of cerebral blood flow in man during arm work. Brain 94: 635–646

– (1972) Paulson OB, Lassen NA (1971) Regional cerebral blood flow in man determined by the initial slope of the clearance of the intra-arterially injected 133Xe. Stroke 2: 519–540

– (1972) The effect of intracarotid epinephrine, norepinephrine and angiotensin on regional cerebral blood flow in man. Neurology (Minneap) 22: 978–987

Ommaya AK, Gennaralli TA (1974) Cerebral concussion and traumatic unconsciousness. Correlation of experimental and clinical observations on blunt head injuries. Brain 97: 633–654

Overgaard J, Tweed WA (1974) Cerebral circulation after head injury. Part I: Cerebral blood flow and its regulation after closed head injury with emphasis on clinical correlations. J Neurosurg 41: 531–541

– (1975) Reflections on prognostic determinants in acute head injury. In: McLaurin RL (ed) Head injuries. Grune and Stratton, New York, pp 11–23

– Tweed WA (1975) rCBF in impending brain death. Acta Neurochir (Wien) 31: 167–175

– – (1976) Cerebral circulation after head injury. Part 2: The effects of traumatic brain edema. J Neurosurg 45: 292–300

– Mosdal C, Tweed WA (1981) Cerebral circulation after head injury. Part 3: Does reduced regional cerebral blood flow determine recovery of brain function after blunt head injury? J Neurosurg 55: 63–74

– Tweed WA (1983) Cerebral circulation after head injury. Part 4: Functional anatomy and boundary-zone flow deprivation in the first week of traumatic coma. J Neurosurg 59: 439–446

Palvölgyi R (1969) Regional cerebral blood flow in patients with intracranial tumors. J Neurosurg 31: 149–163

Papo I, Caruselli G (1977) Long-term intracranial pressure monitoring in comatose patients suffering from head injuries. A critical survey. Acta Neurochir (Wien) 39: 187–200

Pappius HM (1981) Local cerebral glucose utilization in thermally traumatized rat brain. Ann Neurol 9: 484–491

Paulson OB (1970) Regional cerebral blood flow in apoplexy due to occlusion of the middle cerebral artery. Neurology (Minneap) 20: 63–77

– Olesen J, Christensen MS (1972) Restoration of autoregulation of cerebral blood flow by hypocapnia. Neurology (Minneap) 22: 286–293

Pena H, Gaines C, Suess D, Crowell RM, Waggener JD, DeGirolami U (1982) Effect of mannitol on experimental focal ischaemia in awake monkeys. Neurosurgery 11: 477–481

Pfenninger E, Lindner KH, Ahnefeld FW (1989 a) Die Infusion von THAM (trishydroxymethylaminomethan) als Therapie zur Senkung des erhöhten intrakraniellen Druckes beim akuten Schädel-Hirn-Trauma. Anaesthesist 38: 189–192

– Reith A, Breitig D, Grünert A, Ahnefeld FW (1989 b) Early changes of intracranial pressure, perfusion pressure, and blood flow after acute head injury. J Neurosurg 70: 774–779

Phillis JW, De Long RE (1987) An involvement of adenosine in cerebral blood flow regulation during hypercapnia. Gen Pharmac 18: 133–139

Phleps ME, Hoffman EJ, Mullani NA, Ter-Pogossian MM (1975) Application of annihilation coincidence detection to transaxial reconstruction tomography. J Nucl Med 16: 210–223

– Mazziotta JC, Huang SC (1982) Study of cerebral function with positron computed tomography. J Cereb Blood Flow Metab 2: 113–162

Pierce EC, Lambertsen CJ, Deutsch S, Chase PE, Linde HW, Dripps RD, Price HL (1962) Cerebral circulation and metabolism during thiopental anesthesia and hyperventilation in man. J Clin Invest 41: 1664–1671

Pistolese GR, Faraglia V, Agnoli A, Prencipe M, Pastore E, Spartera C, Fiorani P (1972) Cerebral hemispheric "counter-steal" phenomenon during hyperventilation in cerebrovascular diseases. Stroke 3: 456–461

Plum F, Siesjö BK (1975) Recent advances in CSF physiology. Anesthesiology 42: 708–730

Prado R, Ginsberg MD, Dietrich WD, Watson BD, Busto R (1988) Hyperglycemia increases infarct size in collaterally perfused but not end-arterial vascular territories. J Cereb Blood Flow Metab 8: 186–192

Pulsinelli WA, Levy DE, Duffy TE (1982 a) Regional cerebral blood flow and glucose metabolism following transient forebrain ischemia. Ann Neurol 11: 499–509

– Bierley IB, Plum F (1982 b) Temporal profile of neuronal damage in a model of transient forebrain ischemia. Ann Neurol 11: 491–498

– French J, Rawlinson D, Plum F (1982 c) Cerebral ischemia damages neurons despite lowered brain lactate levels. Ann Neurol 12: 86

Rabow L, Hedman G (1979) CKBB-isoenzymes as a sign of cerebral injury. Acta Neurochir (Wien) [Suppl] 28: pp 108–112

– – (1985) Creatine-kinase BB activity after head trauma related to outcome. Acta Neurochir (Wien) 76: 137–139

– DeSalles AAF, Becker DP, Yang M, Kontos HA, Ward JD, Moulton RJ, Clifton G, Gruemer HD, Muizelaar JP, Marmarou A (1986) CSF brain creatine kinase levels and lactic acidosis in severe head injury. J Neurosurg 65: 625–629

Raichle ME, Posner JB, Plum F (1970) Cerebral blood flow during and after hyperventilation. Arch Neurol 23: 394–403

– Plum F (1972) Hyperventilation and cerebral blood flow. Stroke 3: 566–575

– (1983) Positron emission tomography. Ann Rev Neurosci 6: 249–267

Rapela CE, Green HD (1964) Autoregulation of canine cerebral blood flow. Circ Res 15: 205–211

– – Denison AB (1967) Baroreceptor reflexes and autoregulation of cerebral blood flow in the dog. Circ Res 21: 559–568

Rapoport SI, Bachman DS, Thompson HK (1972) Chronic effects of osmotic opening of the blood-brain barrier in the monkey. Science 176: 1243–1245

Ravussin P, Archer DP, Meyer E, Abou-Madi M, Yamamoto L, Trop D (1985) Effects of rapid infusion on cerebral blood volume and intracranial pressure in dogs. Can Anaesth Soc J 32: 506–515

– – Tyler JL, Meyer E, Abou-Madi M, Diksic M, Yamamoto L, Trop D (1986 a) Effects of rapid mannitol infusion on cerebral blood volume. J Neurosurg 64: 104–113

– Chiolero R, Buchser E, DeTribolet N, Freeman J (1986 b) CSF pressure changes following mannitol in patients undergoing craniotomy. Anesthesiology 65: A 303

– Abou-Madi H, Archer D, Chiolero R, Freeman J, Trop D, DeTribolet N (1988) Changes in CSF pressure after mannitol in patients with and without elevated CSF pressure. J Neurosurg 69: 869–876

Reeder RF, Nattie EE, North WG (1986) Effect of vasopressin on cold-induced brain edema in cats. J Neurosurg 64: 941–950

Rehncrona S, Rosen I, Siesjö BK (1981) Brain lactic acidosis and ischaemic cell damage: 1. Biochemistry and neurophysiology. J Cereb Blood Flow Metab 1: 297–311

Reivich M (1964) Arterial PCO$_2$ and cerebral hemodynamics. Am J Physiol 206: 25–35

– Cohen PJ, Greenbaum L (1966) Alterations in the electroencephalogram of awake man produced by hyperventilation: Effects of 100% oxygen at 3 atmospheres (absolute) pressure. Neurology 16: 304

– Jehle J, Sokoloff L, Kety SS (1969) Measurement of regional cerebral blood flow with antipyrine-14C in awake cats. J Appl Physiol 27: 296–300

Reulen HJ, Kreysch HG (1973) Measurement of brain tissue pressure in cold induced cerebral oedema. Acta Neurochir (Wien) 29: 29–40

– Graham R, Spatz M, Klatzo I (1977) Role of pressure gradients and bulk flow in dynamics of vasogenic edema. J Neurosurg 46: 24–35

Reynier-Rebuffel A, Aubineau P, Issertial O, Seylaz J (1987) Non-uniformity of CBF response to NE- or ANG II-induced hypertension in rabbits. Am J Physiol 253: H 47–H 57

Risberg J, Ali Z, Wilson EM, Wills EL, Halsey JH (1975) Regional cerebral blood flow by 133-Xenon inhalation: Preliminary evaluation of an initial slope index in patients with unstable flow compartments. Stroke 6: 142–148

– (1980) Regional cerebral blood flow measurements by 133 xenon inhalation: methodology and application in neuropsychology and psychiatry. Brain Lang 9: 9–34

Robertson CS, Grossman RG, Goodman JC, Narayan RK (1987) The predictive value of cerebral anaerobic metabolism with cerebral infarction after head injury. J Neurosurg 67: 361–368

– Narayan RK, Gokaslan ZL, Pahwa R, Grossman RG, Caram P, Allen E (1989) Cerebral arteriovenous oxygen difference as an estimate of cerebral blood flow in comatose patients. J Neurosurg 70: 222–230

Rockoff MA, Marshall LF, Shapiro HM (1979) High-dose barbiturate therapy in humans: A clinical review of 60 patients. Ann Neurol 6: 194–199

Rosner MJ, Becker DP (1984) Experimental brain injury: successful therapy with a weak base, tromethamine. With an overview of CNS acidosis. J Neurosurg 60: 961–971

– Coley I (1986) Cerebral perfusion pressure, the ICP and head elevation. J Neurosurg 65: 636–641

– – (1987) Cerebral perfusion pressure: A hemodynamic mechanism of mannitol and postmannitol hemogram. Neurosurgery 21: 147–156

– (1987) Cerebral perfusion pressure: Link between intracranial pressure and systemic circulation. In: Wood JH (ed) Cerebral blood flow, physiological and clinical aspects. McGraw-Hill Company, pp 425–448

– Elias KG, Coley I (1989) Prospective, randomized trial of THAM therapy in severe brain injury: Preliminary results. In: Hoff JT, Betz AL (eds) Intracranial pressure VII. Springer, Berlin Heidelberg New York, pp 611–615

Rosomoff HL, Holaday DA (1954) Cerebral blood flow and cerebral oxygen consumption during hypothermia. Am J Physiol 179: 85–88

– (1963) Distribution of intracranial contents with controlled hyperventilation: Implications of neuroanesthesia. Anesthesiology 24: 640–645

Rossi GT, Britt RH (1984) Effects of hypothermia on the cat brainstem auditory evoked response. Electroencephalogr Clin Neurophysiol 57: 143–155

Roth JA, Greenfield AJ, Kaihara S, Wagner HN (1970) Total and regional cerebral blood flow in unanesthetized dogs. Am J Physiol 219: 96–101

Rowan JO, Reilly P, Farrar JK, Teasdale G (1975) The Xenon-133 and hydrogen clearance methods a comparative analysis. In: Harper AM, Jennett WB, Miller JD, Rowan JO (eds) Blood flow and metabolism in the brain. Churchill Livingstone, Edinburgh, pp 7.9–7.10

Rudehill A, Lagerkranser M, Lindquist C, Gordon E (1983) Effects of mannitol on blood volume and central hemodynamics in patients undergoing cerebral aneurysm surgery. Anesth Analg 62: 875–880

Safar P, Stezoski W, Nemoto EM (1976) Amelioration of brain damage after 12 minutes cardiac arrest in dogs. Arch Neurol 33: 91–95

Sakurada O, Kennedy C, Jehle J, Brown JD, Carbin GL, Sokoloff L (1978) Measurement of local cerebral blood flow with iodo (14C) antipyrine. Am J Physiol 234: 59–66

Salanga VD, Waltz AG (1973) Regional cerebral blood flow during stimulation of seventh cranial nerve. Stroke 4: 213–217

Samra SK, Turk P, Arens JF (1989) Effect of hypocapnia on local cerebral glucose utilization in rats. Anesthesiology 70: 523–526

Sapirstein LA, Ogden E (1956) Theoretic limitations of the nitrous oxide method for the determination of regional blood flow. Circ Res 4: 245–249

Sato M, Pawlik G, Heiss WD (1984) Comparative studies of regional CNS blood flow autoregulation and responses to CO_2 in the cat. Effects of altering arterial blood pressure and $PaCO_2$ on rCBF of cerebrum, cerebellum, and the spinal cord. Stroke 15: 91–97

Saul TG, Ducker TB (1982) Effect of intracranial pressure monitoring and aggressive treatment on mortality in severe head injury. J Neurosurg 56: 498–503

Saunders ML, Miller DJ, Stablein D, Allen G (1979) The effects of graded experimental trauma on cerebral blood flow and responsiveness to CO_2. J Neurosurg 51: 18–26

Sawada Y, Sugimoto H, Kobayashi H, Ohashi N, Yoshioka T, Sugimoto T (1982) Acute tolerance to high-dose barbiturate treatment in patients with severe head injuries. Anesthesiology 56: 53–54

Schutta HS, Kassell NF, Langfitt TW (1968) Brain swelling produced by injury and aggravated by arterial hypertension. A light and electronic microscopic study. Brain 91: 281–292

Sedzimir CB (1959) Therapeutic hypothermia in cases of head injury. J Neurosurg 16: 407–414

Seelig JM, Lewelt W, Jenkins JD, Miller JD, Becker DP (1983) Autoregulation of CBF in response to changes in arterial and intracranial pressure after experimental head injury. In: Ishii S, Nagai H, Brock M (eds) Intracranial pressure V. Springer, Berlin Heidelberg New York, pp 487–489

Segawa H, Wakai S, Tamura A, Yoshimasu N, Nakamura O, Ohta M (1983) Computed tomographic measurement of local cerebral blood flow by Xenon enhancement. Stroke 14: 356–362

Seitz HD, Ocker K (1977) The prognostic and therapeutic importance of changes in the CSF during the acute stage of brain injury. Acta Neurochir (Wien) 38: 211–231

Seki H, Ogawa A, Yoshimoto T, Suzuki J (1981) Effect of mannitol on rCBF in canine thalamic ischemia. An experimental study. No To Shinkei 33: 1101–1105

Selman RW, Spetzler RF, Roessmann UR, Rosenblatt JI, Crumrine RC (1981) Barbiturate infusion coma therapy for focal cerebral ischaemia. J Neurosurg 55: 220–226

Sercombe R, Verrechia C, Oudart N, Dimitriadou V, Seylaz J (1985) Pial artery responses to norepinephrine potentiated by endothelium removal. J Cereb Blood Flow Metab 5: 312–317

Shapiro HM, Wyte SR, Loeser J (1974) Barbiturate-augmented hypothermia for reduction of persistant intracranial hypertension. J Neurosurg 40: 90–100

– (1985) Barbiturates in brain ischaemia. Br J Anaesth 57: 82–95

Shalit MN, Beller AJ, Feinsod M (1972) Clinical equivalents of cerebral oxygen consumption in coma. Neurology 22: 155–160

Sharbrough FW, Messick JM, Sundt TM Jr (1973) Correlation of continuous electroencephalograms with cerebral blood flow measurements during carotid endarterectomy. Stroke 4: 674–683

Shenkin HA, Hafkenschiel JH, Kety SS (1950) Effects of sympathectomy on the cerebral circulation of hypertensive patients. Arch Surg 61: 319–324

– Goluboff B, Haft H (1962) The use of mannitol for the reduction of intracranial pressure in intracranial surgery. J Neurosurg 19: 897–901

Shohami E, Shapira Y, Sidi A, Cotev S (1987) Head injury induces increased prostaglandin synthesis in rat brain. J Cereb Blood Flow Metab 7: 58–63

Sidi A, Cotev S, Hadani M, Wald U, Feinsod M, Perel A (1983) Long-term barbiturate infusion to reduce intracranial pressure. Crit Care Med 11: 478–481

Siemkowicz E, Hansen A (1978) Clinical restitution following cerebral ischemia in hypo-, normo-, and hyperglycemic rats. Acta Neurol Scand 58: 1–8

– – (1981) Brain extracellular ion composition and EEG activity following 10 min ischemia in normo- and hyperglycemic rats. Stroke 12: 236–240

Siesjö BK, Ingvar M, Pelligrino D (1983) Regional diferences in vascular autoregulation in the rat brain in severe insulin-induced hypoglycemia. J Cereb Blood Flow Metab 3: 478–485

– (1984) Cerebral circulation and metabolism. J Neurosurg 60: 883–908

– Wieloch T (1985) Brain injury: neurochemical aspects. In: Becker C, Povlbshock J (eds) Central nervous system trauma status report. William Byrd Press, New York, pp 513–532

– (1988) Mechanisms of ischaemic brain damage. Crit Care Med 16: 954–963

Skinhøj E, Strandgaard S (1973) Pathogenesis of hypertensive encephalopathy. Lancet i: 461–462

Slocum HC, Hayes GW, Laezman BL (1961) Ventilator technique of anesthesia for neurosurgery. Anesthesiology 22: 143–145

Smith AL, Hoff JT, Nielsen SL, Larson CP (1974) Barbiturate protection in acute focal cerebral ischaemia. Stroke 5: 1–7

Smith DS, Rehncrona S, Siesjö BK (1980) Inhibitory effects of different barbiturates on lipid peroxidation in the brain tissue in vitro: Comparison with the effects of promethazine and chlorpromazine. Anesthesiology 53: 186–194

Snyder BD, Ramirez-Lassepas M, Sukhum P, Fryd D, Sung JH (1979) Failure of thiopental to modify global anoxic injury. Stroke 10: 135–141

Snyder JV, Nemoto EM, Carroll RG, Safar P (1975) Global ischemia in dogs. Intracranial pressures, brain blood flow and metabolism. Stroke 6: 21–27

Sokoloff L, Reivich M, Kennedy C, DesRosiers MH, Patlak CS, Pettigrew KD, Sakurada O, Shinohara M (1977) The (14C) deoxyglucose method for the measurement of local cerebral glucose utilization: Theory, procedure, and normal values in the conscious and anesthetized albino rat. J Neurochem 28: 897–916

Soloway M, Nadel W, Albin MS, White RJ (1968) The effect of hyperventilation on subsequent cerebral infarction. Anesthesiology 29: 975–980

– Moriarty G, Fraser JG, White RJ (1971) Effect of delayed hyperventilation on experimental cerebral infarction. Neurology 21: 479–485

Sood SC, Gulati SC, Kumar M, Kak VK (1980) Cerebral metabolism following brain injury II. Lactic acid changes. Acta Neurochir (Wien) 53: 47–51

Speth RC, Harik SI (1985) Angiotensin II receptor binding sites in brain microvessels. Proc Natl Sci 82: 6340–6343

Spetzler RF, Selman WR, Roski RA, Bonstelle C (1982) Cerebral revascularization during barbiturate coma in primates and humans. Surg Neurol 17: 111–115

Steen PA, Milde JH, Michenfelder JD (1978) Cerebral metabolic and vascular effects of barbiturate therapy following complete global ischemia. J Neurochem 31: 1317–1324

– – – (1979) No barbiturate protection in a dog model of complete cerebral ischemia. Ann Neurol 5: 343–349

Steen PA, Newberg L, Milde JH, Michenfelder JD (1983) Hypothermia and barbiturates: Individual and combined effect on canine cerebra oxygen consumption. Anesthesiology 58: 527–532

Steinbach JJ, Mattar AG, Mahin DT (1976) Alteration of the cerebral blood flow study due to reflux in internal jugular veins. J Nucl Med 17: 61–64

Stokely EM, Sveinsdottir E, Lassen NA, Rommer P (1980) A single photon dynamic computer assisted tomography (DCAT) for imaging brain function in multiple cross sections. J Comput Assist Tomogr 4: 230–240

Strandgaard S, Olesen J, Skinhøj E, Lassen NA (1973) Autoregulation of brain circulation in severe arterial hypertension. Br Med J 1: 507–510

– McKenzie ET, Jones JV, Harper AM (1975 a) Studies on cerebral blood flow following breakthrough of autoregulation. In: Harper M, Jennett B, Miller D, Rowan J (eds) Blood flow and metabolism in the brain. Churchill Livingstone, Edinburgh London New York, pp 5.15

– Jones JV, MacKenzie ET, Harper AM (1975 b) Upper-limit of cerebral blood flow autoregulation in experimental renovascular hypertension in the baboon. Circ Res 37: 164–167

– (1976) Autoregulation of cerebral blood flow in hypertensive patients. The modifying influence of prolonged antihypertensive treatment on the tolerance to acute, drug-induced hypotension. Circulation 53: 720–727

– Paulson OB (1984) Cerebral autoregulation. Stroke 15: 413–416

Strong AJ, Venables GS, Gibson G (1983) The cortical ischemic penumbra associated with occlusion of MCA in the cat. Topography of changes in blood flow, potassium ion activity, and EEG. J Cereb Blood Flow Metab 3: 86–96

– Gibson G, Miller SA, Venables GS (1988) Changes in vascular and metabolic reactivity as indices of ischaemia in the penumbra. J Cereb Blood Flow Metab 8: 79–88

Stullken AH, Milde JH, Michenfelder JD, Tinker JH (1977) The nonlinear responses of cerebral metabolism to low concentration of halothane, enflurane, isoflurane and thiopental. Anesthesiology 46: 28–34

Sullivan HG, Martinez J, Becker DP et al (1976) Fluid percussion model of mechanical brain injury in the cat. J Neurosurg 45: 520–534

– Miller DJ, Becker DP, Flora RE, Allen DA (1977) The physiological basis of intracranial pressure change with progressive epidural brain compression. J Neurosurg 47: 532–550

Sundbärg G (1988) Neurosurgical intensive care and the management of severe head injuries. Lund, Studentlitteratur (Thesis)

Sutherland G, Lesiuk H, Bose R, Sima AAF (1988) Effect of mannitol, nimodipine, and indomethacin singly or in combination on cerebral ischaemia in rats. Stroke 19: 571–578

Sutton LN, McLaughlin AC, Kemp W, Schnall MD, Cho B, Langfitt TW, Chance B (1987) Effect of increased ICP on brain phosphocreatine and lactate determined by simultaneous 1H and 31P NMR spectroscopy. J Neurosurg 67: 381–386

Suzuki M, Iwasaki Y, Yamamoto T, Konno H, Kudo H (1988) Sequelae of the osmotic blood-brain barrier opening in rats. J Neurosurg 69: 421–428

Suzuki R, Nitsch C, Fujiwara K, Klatzo I (1984) Regional changes in cerebral blood flow and blood-brain barrier permeability during epileptiform seizures and in acute hypertension in rabbits. J Cereb Blood Flow Metab 4: 96–102

Sveinsdottir E, Larsen B, Rommer P, Lassen NA (1977) A multidetector scintillation camera with 254 channels. J Nucl Med 18: 168–174

Symon L, Held K, Dorsch MWC (1972) On the myogenic nature of the autoregulatory mechanism in the cerebral circulation. Europ Neurol 6: 11–18

– – – (1973) A study of regional autoregulation in the cerebral circulation to increased perfusion pessure in normocapnia and hypercapnia. Stroke 4: 139–147

– Brierley J (1976) Morphological changes in cerebral blood vessels in chronic ischaemic infarction: Flow correlation obtained by the hydrogen clearance method. In: Cervos-Navarro J, Matakas F et al (eds) The cerebral vessel wall. Raven Press, New York, pp 165–174

– Branston NM, Chikovani O (1979) Ischemic brain edema following middle cerebral artery occlusion in baboons: Relationship between regional cerebral water content and blood flow at 1 and 2 hours. Stroke 10: 184–191

– (1980) The relationship between CBF, evoked potentials, and the clinical features in cerebral ischemia. Acta Neurol Scand 62 [Suppl] 78: 175–190

Søndergård W (1961) Intracranial pressure during general anaesthesia. Dan med Bull 8: 18–26

Sørensen SC (1978) Theoretical considerations on the potential hazards of hyperventilation during anaesthesia. Acta Anaesthesiol Scand [Suppl] 67: 106–110

Tabaddor K, Gardner TJ, Walker AE (1972) Cerebral circulation and metabolism at deep hypothermia. Neurology (Minneap) 22: 1065–1070

– Bhushan C, Pevsner PH, Walker AE (1972) Prognostic value of cerebral blood flow (CBF) and cerebral metabolic rate of oxygen (CMRO$_2$) in acute head trauma. J Trauma 12: 1053–1055

Tanaka A, Tomonaga M (1987) Effect of mannitol on cerebral blood flow and microcirculation during experimental middle cerebral artery occlusion. Surg Neurol 28: 189–195

Tans JTJ, Poortvliet DCJ (1983) Intracranial volume-pressure relationship in man. Part 2: Clinical significance of the pressure volume index. J Neurosurg 59: 810–816

Teasdale G, Jennet B (1976) Assessment and prognosis of coma after head injury. Acta Neurochir (Wien) 34: 45–55

Teasdale E, Cardoso E, Galbraith S, Teasdale G (1984) CT scan in severe diffuse head injury: physiological and clinical correlations. J Neurol Neurosurg Psychiatry 47: 600–603

Teasdale GM, Mendelow AD, Galbraith S (1986) Causes and consequences of raised intracranial pressure in head injuries. In: Miller JD, Teasdale GM, Rowan JO, Galbraith SL, Mendelow AD (eds) Intracranial pressure VI. Springer, Berlin Heidelberg New York, pp 3–8

Tenjin H, Mizukawa N, Yamaki T, Imahori Y, Hino A, Hirakawa K, Yamashita M, Nakahashi H (1989) The investigation of cerebral hemodynamics in patients with cerebral contusion: Evaluation by positron emission tomography (PET). J Cereb Blood Flow Metab [Suppl] 1: S 87

Ter-Pogossian MM (1977) Basic priniciples of computed axial to-mography. Semin Nucl Med 7: 109–127

Terry HR, Daw EF, Michenfelder JD (1962) Hypothermia by ex-tracorporeal circulation for neurosurgery: An anesthetic tech-nique. Anesth Analg 41: 241–248

Thorn W, Heitmann R (1954) pH der Gerhirnrinde vom Kaninchen in Situ während perakuter, totaler Ischaemie, reiner Anoxie und in der Erholung. Pflügers Arch 258: 501–510

Todd MM, Chadwick HS, Shapiro HM, Dunlop BJ, Marshall LF, Dueck R (1982) The neurologic effects of thiopental therapy following experimental cardiac arrest in cats. Anesthesiology 57: 76–86

Tominaga S, Strandgaard S, Uemera K, Ito K, Kutsuzawa T, Lassen NA, Nakamura T (1976) Cerebrovascular CO_2 reactivity in nor-motensive and hypertensive man. Stroke 7: 507–510

Tomida S, Nowak TS, Vass K, Lohr JM, Klatzko I (1987) Exper-imental model for repetitive ischemic attacks in the gerbil: The cumulative effect of repeated ischaemic insults. J Cereb Blood Flow Metab 7: 773–782

Tomita M, Gotoh F (1981) Local cerebral blood flow values as estimated with diffusible tracers: Validity of assumptions in nor-mal and ischemic tissue. J Cereb Blood Flow Metab 1: 403–411

Tomiwa K, Hazama F, Mikawa H (1982) Reversible osmotic opening of the blood-brain barrier. Prevention of tissue damage with filtration of the perfusate. Acta Pathol Jpn 32: 427–435

Tornheim PA, McLaurin RL (1981) Acute changes in regional brain water content following experimental closed head injury. J Neu-rosurg 55: 407–413

– Liwnicz BH, Hirsch CS, Brown DL, McLaurin RL (1983) Acute responses to blunt head trauma. Experimental model and gross pathology. J Neurosurg 59: 431–438

Troupp H (1967) Intraventricular pressure in patients with severe brain injuries. J Trauma 7: 875–883

– Vapalahti M (1971) Intraventricular pressure in the final stages of severe brain injury. Acta Neurochir (Wien) 25: 189–195

Trøjaborg W, Boysen G (1973) Relation between EEG, regional cerebral blood flow and internal carotid artery pressure during carotid endarterectomy. EEG Clin Neurophysiol 34: 61–69

Tuor UI, Farrar JK (1984) Pial vessel caliber and cerebral blood flow during hemorrhage and hypercapnia in the rabbit. Am J Physiol 247: H 40–51

Uihlein A, Terry HR, Payne WS, Kirklin JW (1962) Operations on intracranial aneurysms with induced hypothermia below 15 °C and total circulatory arrest. J Neurosurg 19: 237–239

– MacCarty CS, Michenfelder JD, Terry HR, Daw EF (1966) Deep hypothermia and surgical treatment of intracranial aneurysms. JAMA 195: 639–641

Unterberg A, Baethmann AJ (1984) The kallikrein-kinin system as mediator in vasogenic brain edema. Part 1: Cerebral exposure to bradykinin and plasma. J Neurosurg 61: 87–96

– Wahl M, Baethmann A (1984) Effects of bradykinin on per-meability and diameter of pial vessels in vivo. J Cereb Blood Flow Metab 4: 574–585

– Dautermann C, Baethmann A, Muller-Esterl W (1986) The kal-likrein-kinin system as mediator in vasogenic brain edema. J Neurosurg 64: 269–276

– Andersen BJ, Clarke GD, Marmarou A (1988) Cerebral energy metabolism following fluid percussion brain injury in cats. J Neurosurg 68: 594–600

Uzzell BP, Obrist WD, Dolinskas CA, Langfitt TW (1986) Rela-tionship of acute CBF and ICP findings to neuropsychological outcome in severe head injury. J Neurosurg 65: 630–635

Vanhoutte PM, Verbeuren TJ, Webb RC (1981) Local modulation of adrenergic neuroeffector interaction in the blood vessel wall. Physiol Rev 61: 151–247

Van Wylen DGL, Park TS, Rubio R, Berne RM (1987) Brain di-alysate adenosine concentration during cerebral autoregulation in the adult rat. Fed Proc 46: 354

Vapalahti M, Troupp H, Heiskanen O (1969) Extremely severe brain injuries treated with hyperventilation and ventricular drainage. In: Brock M, Fieschi C, Ingvar DH, Lassen NA, Schurmann K (eds) Cerebral blood flow. Springer, Berlin Heidelberg New York, pp266–267

Vinall PE, Simeone FA (1981) Cerebral autoregulation: An in vitro study. Stroke 12: 640–642

Vink R, McIntosh TK, Weiner MW, Faden AI (1987) Effects of traumatic brain injury on cerebral high-energy phosphates and pH: A 31P magnetic resonance spectroscopy study. J Cereb Blood Flow Metab 7: 563–571

Wagner KR, Tronheim PA, Eichhold MK (1985) Acute changes in regional cerebral metabolite values following experimental blunt head injury. J Neurosurg 63: 88–96

Wahl M, Young AR, Edvinsson L, Wagner F (1983) Effects of bradykinin on pial arteries and arterioles in vitro and in situ. J Cereb Blood Flow Metab 3: 231–237

Waltz AG, Yamaguchi T, Regli F (1971) Regulatory responses of cerebral vasculature after sympathetic denervation. Am J Phy-siol 221: 298–302

Ward JD, Becker DP, Miller JD, Choi SC, Marmarou A, Wood C, Newlon PG, Keenan R (1985) Failure of prophylactic barbi-turate coma in the treatment of severe head injury. J Neurosurg 62: 383–388

– Choi S, Marmarou A, Moulton R, Muizelaar JP, DeSalles A, Becker DP, Kontos HA, Young HF (1989) Effect of prophy-lactic hyperventilation on outcome in patients with severe head injury. In: Hoff JT, Betz AL (eds) Intracranial pressure VII. Springer, Berlin Heidelberg New York, pp 630–633

Warner DS, Turner DM, Kassell NF (1987) Time-dependent effects of prolonged hypercapnia on cerebrovascular parameters in dogs: Acid-base chemistry. Stroke 18: 142–149

Watanaba T, Yoshimoto T, Ogawa A, Sakamoto T, Suzuki J (1979) The effect of mannitol in preventing the development of cerebral infarction. An electron microscopic investigation. No Shinkei Geka 7: 859–866

Wechsler RL, Dripps RD, Kety SS (1951) Blood flow and oxygen consumption of the human brain during anesthesia produced by thiopental. Anesthesiology 12: 308–313

Wei EP, Ellis EF, Kontos HA (1980) Role of prostaglandins in pial arteriolar response to CO_2 and hypoxia. Am J Physiol 238: H 226–230

– Lamp RG, Kontos HA (1982) Increased phospholipase C activity after experimental brain injury. J Neurosurg 56: 695–698

White RJ, Albin MS, Verdura J, Locke GE (1967) Differential ex-tracorporal hypothermic perfusion of and circulatory arrest to the human brain. Med Res Engineering 6: 18–24

– (1972) Preservation of cerebral function during circulatory arrest and resuscitation: Hypothermic protective considerations. Re-suscitation 1: 107–115

Winkler S, Holden J, Sackett JF, et al (1977) Xenon inhalation as an adjunct to computerized tomography of the brain: Prelimi-nary study. Invest 12: 15–18

Winn HR, Welsh JE, Rubio R, Berne RM (1980) Brain adenosine production in the rat during sustained alteration in systemic blood pressure. Am J Physiol 239: 636–641

– Rubio R, Berne RM (1981) The role of adenosine in the regulation of cerebral blood flow (editorial). J Cereb Blood Flow Metab 1: 239–244

Wise BL, Chater N (1962) The value of hypertonic mannitol solution in decreasing brain mass and lowering cerebrospinal-fluid pressure. J Neurosurg 19: 1038–1043

Wollman SB, Orkim LR (1968) Postoperative human reaction time and hypocapnia during anaesthesia. Br J Anaesth 40: 920–927

Wozney P, Yonas H, Latchaw RE, Gur D, Good W (1985) Central herniation revealed by focal decrease in blood flow without elevation of intracranial pressure.: A case report. Neurosurgery 17: 641–644

Wyper DJ, Brooke MBD (1977) Compensating for hemisphere crosstalk when measuring CBF. Acta Neurol Scand 56 [Suppl] 64: 470–471

Yang M, DeWitts DS, Becker DP, Hayes RL (1985) Regional brain metabolite levels following mild experimental head injury in the cat. J Neurosurg 63: 617–621

Yano M, Ikeda Y, Kobayshi S, Yamamoto A, Otsuka T (1986) The outcome with barbiturate therapy in severe head injuries. In: Miller JD, Teasdale GM, Rowan JO, Galbraith SL, Mendelow AD (eds) Intracranial pressure VI. Springer, Berlin Heidelberg New York, pp 769–773

– – – Otsuka T (1987) Intracranial pressure in head-injured patients with various intracranial lesions is identical throughout the supratentorial intracranial compartment. Neurosurgery 21: 688–692

Yonas H, Dujovny M, Nelson D, Lipton SD, Segel R, Agdeppa D, Mazel M (1981) The controlled delivery of thiopental and delayed cerebral revascularization. Surg Neurol 15: 27–34

– Wolfson SK, Gur D, Latchaw RE, Good WF, Leanza R, Jackson DL, Jannetta PJ, Reinmuth OM (1984 a) Clinical experience with the use of xenon-enhanced CT blood blow mapping in cerebral vascular disease. Stroke 15: 443–450

– Good WF, Gur D, Wolfson SK Jr, Latchaw RE, Good BC, Leanza R, Miller SL (1984 b) Mapping cerebral blood flow by xenon-enhanced computed tomography: Clinical experience. Radiology 152: 425–442

– Snyder JV, Gur D, Good WF, Latchaw RE, Wolfson SK, Grevik A, Good BC (1984 c) Local cerebral blood flow alterations (Xe-CT method) in an accident victim. J Comput Assist Tomogr 8: 990–991

– Gur D, Good BC et al (1985) Stable xenon CT blood flow mapping for evaluation of patients with extracranial/intracranial bypass surgery. J Neurosurg 62: 324–333

– – Latchaw RE, Wolfson SK (1987) Xenon computed tomographic blood flow mapping. In: Wood JH (ed) Cerebral blood flow. McGraw-Hill Book Company, pp 220–242

– – Claassen D, Wolfson SK, Moossy J (1988) Stable Xenon enhanced computed tomography in the study of clinical and pathologic correlates of focal ischemia in baboons. Stroke 19: 228–238

Young RS, Olenginski TP, Yagel SK, Towfighi J (1983) The effect of graded hypothermia on hypoxic-ischemic brain damage: A neuropathologic study in the neonatal rat. Stroke 14: 929–934

Zasslow MA, Pearl RG, Shuer LM, Lieberson RE, Steinberg GK, Larson CP (1987) Hyperglycemia decreases neuronal ischemic changes after middle cerebral artery occlusion in the cat. Anesthesiology 67: A 581

Zupping R (1970) Cerebral acid-base and gas metabolism in brain injury. J Neurosurg 33: 498–505

Zwetnow NN (1970) The influence of an increased intracranial pressure on the lactate, pyruvate, bicarbonate, phosphocreeatine, ATP, ADP and AMP concentrations of the cerebral cortex of dogs. Acta Physiol Scand [Suppl] 339: 1–31

New by Springer-Verlag

John D. Pickard, François Cohadon, João Lobo Antunes (Eds.)

Neuroendocrinological Aspects of Neurosurgery

Proceedings of the Third Advanced Seminar in Neurological Research, Venice, April 30–May 1, 1987

Acta Neurochirurgica / Supplementum 47

Contents: B. J. Everitt and T. Hökfelt: Neuroendocrine Anatomy of the Hypothalamus. – J. D. Vincent and G. Simonnet: Neurohormonal Communication in the Brain. – I. Assenmacher: Central Control of Circadian and Ultradian Neuroendocrine Rhythms. – J. Lobo Antunes and K. Muraszko: The Vascular Supply of the Hypothalamus-Pituitary Axis. – M. D. Page et al.: A Clinical Update on Hypothalamic-Pituitary Control. – R. Fahlbusch et al.: Clinical Syndromes of the Hypothalamus. – G. Teasdale: Pituitary Tumours: Problems and Questions. – P. Lees: Intrasellar Pressure. – I. Lancranjan: The Medial Treatment of Prolactin and Growth Hormone-Secreting Pituitary Tumours. – R. M. Buijs: Vasopressin and Oxytocin Localization and Putative Functions in the Brain. – St. Lightman: Central Nervous System Control of Fluid Balance: Physiology and Pathology. – V. Walker: Fluid Balance Disturbances in Neurosurgical Patients: Physiological Basis and Definitions. – G. Neil-Dwyer et al.: The Stress Response in Subarachnoid Haemorrhage and Head Injury. – E. F. M. Wijdicks et al.: Hyponatremia and Volume Status in Aneurysmal Subarachnoid Haemorrhage. – R. J. Nelson: Blood Volume Measurement Following Subarachnoid Haemorrhage. – T. Dóczi et al.: Central Neuroendocrine Control of the Brain Water, Electrolyte, and Volume Homeostasis. – M. Brock: The Hypothalamus: New Ideas on an Old Structure.

1990. 60 figures. VII, 128 pages.
Cloth DM 158,–, öS 1106,–
Reduced price for subscribers
to "Acta Neurochirurgica":
Cloth DM 142,–, öS 995,–
ISBN 3-211-82160-0

Springer-Verlag Wien New York
Moelkerbastei 5, P.O. Box 367, A-1011 Wien
Heidelberger Platz 3, D-1000 Berlin 33
175 Fifth Avenue, New York, NY 10010, USA
37-3, Hongo 3-chome, Bunkyo-ku, Tokyo 113, Japan

Jörn Bo Madsen, Georg Emil Cold

The Effects of Anaesthetics upon Cerebral Circulation and Metabolism

**1990. 14 figures. Approx. 150 pages.
Cloth DM 89,–, öS 623,–
ISBN 3-211-82198-8**

During the last decade, the effects of anaesthetics on cerebral blood flow, cerebral metabolic rate of oxygen and intracranial pressure have been studied experimentally and clinically. In this review studies of CBF and CMRO2 during craniotomy have been performed with the classical technique described by Kety and Schmidt.

In chapter 1 general considerations concerning the effects of anaesthetics on cerebral blood flow and metabolism are reviewed. In chapters 2 and 3, the effects of inhalation agents and hypnotics on flow and metabolism are considered. Chapters 4 and 5 cover the effects of central analgetics, and neuromuscular blocking agents. In chapter 6, the effects of other drugs in common use in neuroanaesthetic practice are summarized. Chapter 7 considers the effects of drugs used for controlled hypotension. In chapter 8, the application of Kety's method in studies of CBF and metabolism is reviewed, the studies of cerebral circulation and metabolism during nine different techniques of anaesthesia for craniotomy are presented, and other studies of cerebral circulation during neuroanaesthesia are reviewed. In chapter 9, considerations concerning central and cerebral hemodynamics during anaesthesia in the sitting position are considered.

This review, is primarily addressed to anaesthetists, but it will also be of interest to those working within neurosurgery, neuroradiology and clinical neurophysiology.

Springer-Verlag Wien New York
Moelkerbastei 5, P.O. Box 367, A-1011 Wien
Heidelberger Platz 3, D-1000 Berlin 33
175 Fifth Avenue, New York, NY 10010, USA
37-3, Hongo 3-chome, Bunkyo-ku, Tokyo 113, Japan